Drone Development from Concept to Flight

Design, assemble, and discover the applications of unmanned aerial vehicles

Sumit Sharma

Drone Development from Concept to Flight

Copyright © 2024 Packt Publishing

All rights reserved. No part of this book may be reproduced, stored in a retrieval system, or transmitted in any form or by any means, without the prior written permission of the publisher, except in the case of brief quotations embedded in critical articles or reviews.

Every effort has been made in the preparation of this book to ensure the accuracy of the information presented. However, the information contained in this book is sold without warranty, either express or implied. Neither the author, nor Packt Publishing or its dealers and distributors, will be held liable for any damages caused or alleged to have been caused directly or indirectly by this book.

Packt Publishing has endeavored to provide trademark information about all of the companies and products mentioned in this book by the appropriate use of capitals. However, Packt Publishing cannot guarantee the accuracy of this information.

Group Product Manager: Preet Ahuja
Publishing Product Manager: Surbhi Suman
Book Project Manager: Uma Devi
Senior Editor: Athikho Sapuni Rishana
Technical Editor: Arjun Varma
Copy Editor: Safis Editing
Proofreader: Athikho Sapuni Rishana
Indexer: Subalakshmi Govindhan
Production Designer: Joshua Misquitta
DevRel Marketing Coordinator: Rohan Dobhal

First published: May 2024
Production reference: 1120424

Published by Packt Publishing Ltd.
Grosvenor House
11 St Paul's Square
Birmingham
B3 1RB, UK.

ISBN 978-1-83763-300-5
www.packtpub.com

To my parents and family, for their support and motivation and for exemplifying the power of determination. To my teachers and mentors, for being my torchlight and guide throughout the journey of life.

– Sumit Sharma

Contributors

About the author

Sumit Sharma is experienced in unmanned systems with a specialization in drone architecture development and testing for different applications. He is experienced in drone mechanics, avionics, design, and manufacturing. In his early career, he was involved in the development and testing of the first approved agriculture drone and later getting it approved by the DGCA. Along with this, he has contributed to the development of a survey drone for aerial surveys. He was also involved in the development of high-speed and long-range surveillance drones for defense tenders. He has been involved in early agriculture spraying pilot projects and was a part of the project that aerial sprayed the locusts that hit Rajasthan in 2020.

I want to thank my parents, mentors, colleagues, and subordinates who have worked closely with me and always supported me to grow.

About the reviewers

Justin Starr, Ph.D., is the endowed professor of advanced technology and the mechatronics program coordinator at the **Community College of Allegheny County (CCAC)**. He works to integrate advanced technology into CCAC's course offerings, including augmented reality, unmanned aerial vehicles, electron microscopy, and collaborative robots. Previously, Dr. Starr served as CTO of RedZone Robotics, a manufacturer of water and wastewater inspection robots. He holds 14 U.S. patents for inventions in robotics, artificial intelligence, and automation. Justin earned his A.S. in general studies from CCAC, his B.S. in engineering science from the University of Virginia, and his M.S. and Ph.D. in materials engineering from the University of Florida.

Garvit Pandya is the founder of the drone manufacturing firm Innovative Unmanned Systems and a qualified engineer with a master's degree in aeronautical engineering with a specialization in unmanned aerial vehicles. He has been working on drones for the last 7+ years, during which he has worked on various projects for army and DRDO units (ADRDE, Agra, and CHESS, Hyderabad) and conducted more than 250 fixed-wing flight operations for a DRDO unit. He has developed various UAV systems for defense and service sector requirements. He has authored a book on drones titled *Basics of Unmanned Aerial Vehicles*, which is available worldwide. The book is listed as a reference book on the curriculum for the Ph.D. in drone technology program at IIT, Guwahati, India.

I am thankful to all the friends, colleagues, and other connections that were directly or indirectly, knowingly or unknowingly, involved in the reviewing process. May God richly bless you all!

I am also thankful to my whole family for all their support, tolerating my busy schedule, and sticking by my side.

Ravi S Shukla is the project manager of a defense electronic manufacturing firm named Alpha Design Technologies Pvt Ltd. His last job was as a technical manager and project associate for IoTechWorld Avigation and IIT Roorkee. He has been active in the UAV industry since 2017. He has rich experience in all types of UAS, such as copter, plane, and rover. Ravi was also involved in forming drone regulations in India. He has great experience in managing different types of projects, such as R&D, design and development, and the full line of production.

I truly believe all of us in the technical world are standing on the shoulders of giants. The giants for me are the open communities, such as CubePilot, ArduPilot, Discord, Mission Planner, and QGCS, where access to information is unrestricted and people are interested in helping one another. I am deeply indebted to all the communities and the people running them. I am also thankful to my whole family for their support.

Table of Contents

Preface — xiii

Part 1: Fundamentals of Flight Engineering

1

Getting Started with UAV and Drone Engineering — 3

Introduction to unmanned systems – unmanned ground, air, and water vehicles — 3	Major mechanical and structural components of a drone — 14
Various unmanned vehicles — 4	Airframe — 14
History and evolution of drones — 4	Avionics systems and subsystems of drones — 17
Need for an unmanned system — 5	The propulsion system or drive train of a drone — 18
What are unmanned aerial systems? — 5	The power system of drones — 20
Types of drones and their relevance to applications — 7	Command and control system — 22
Types of drones and their specifications — 7	Summary — 27
System composition of a UAV — 13	

2

Understanding Flight Terminologies and the Physics of Motion — 29

Introduction to thrust, weight, lift, drag, and cruise — 30	Cruise — 31
Thrust — 30	Introduction to the flight axis and its terminologies — 31
Weight — 30	Roll movement — 32
Lift — 31	Pitch movement — 32
Drag — 31	Yaw movement — 33

Throttle	33	Hover motion	37
Thrust-to-weight ratio (TWR)	34	Horizontal movements	38
Introduction to the flight physics of a quadcopter	34	Pitch movement or forward-backward movement	39
Engineering to achieve movements across all three axes of rotation	37	Summary	40

3

Learning and Applying Basic Command and Control Interfaces   41

Introduction to GCS hardware and software	42	Components of the RC transmitter system	49
GCS hardware	42	Modes of working of an RC controller	50
GCS software	42	Major RC protocols	51
Types of GCS	43	Transmitter protocols	52
Introduction to MAVLink and other open source telemetry protocols	47	Receiving protocols	53
		SBUS protocol	55
Introduction to RCs and the different available options	48	Summary	56

4

Knowing UAV Systems, Sub-Systems, and Components   57

Technical requirements	57	Developing the control system	68
Understanding and conceptualizing the system	58	Electronic speed controllers and flight controllers	68
Designing the propulsion system	59	Flight controller	70
Selection criteria for motor and propeller combination	60	Selecting the navigation and communication systems	73
Characteristics of motors to be used	61	Communication systems	73
Motor specifications and thrust charts	62	Integrated GCS system with communicating devices	75
Motor and propeller selection	65	Selection of GPS and navigation system	76
Battery selection	66	Summary	78
Calculating the flight time (endurance)	67		
Designing and developing the airframe	67		

Part 2: System Conceptualization and Avionics Development

5

The Application of Sensors and IMUs in Drones   81

The inertial measurement unit and its role	81	How does GPS work?	87
		The use of GPS in drones	88
Composition of an IMU	82	Voltage and current sensors and their roles	88
Use of IMUs in unmanned systems	85		
The barometer and its role	86	Sensor fusion and state estimation	89
How does it work?	86	Summary	90
GPS and magnetometer and their roles	87		

6

Introduction to Drone Firmware and Flight Stacks   91

Technical requirements	92	The structure of drone firmware	99
Introduction to firmware	92	Introduction to drone flight stacks	101
Components of firmware and their roles	93	Open source drone flight stack	101
Firmware example	95	Closed source (proprietary) drone flight stack	102
Tools used to develop firmware	95	Basic PX4 controller loop diagram	105
Differences between software and firmware	97	PX4 flight stack	107
Introduction to drone firmware	98	The ArduPilot flight stack	112
How it is different from general firmware	98	Summary	117

7

Introduction to Ground Control Station Software — 119

Technical requirements	120	AeroGCS home menu	129
Introduction to GCSs	120	Methods of connecting a drone	134
Major GCS software that is available on the market	121	RPA or drone configuration	139
		Autonomous mission planning	142
Mission Planner or APM Planner	121	Waypoint planning	143
QGroundControl	123	Survey planning	144
UgCS	124		
AeroGCS software overview	127	Summary	145
AeroGCS main dashboard	127		

8

Understanding Flight Modes and Mission Planning — 147

Technical requirements	147	Configuring the flight modes of the remote controller switch	154
Flight modes and types	148		
Types of fight modes	148	Planning and executing an autonomous waypoint mission	157
Assigning flight modes via remote control	152	Summary	165
How it works	153		
Assigning and changing flight modes in a GCS	153		

Part 3: Configuration, Calibrations, Flying, and Log Analysis

9

Drone Assembly, Configuration, and Tuning — 169

Technical requirements	170	Assembling the components of a drone	173
Components and tools required to build a drone	171	Airframe assembly	173

Motor assembly	175	Firmware flashing	185
Assembling and configuring the ESCs	177	Configuring the flight controller	186
Installing the flight controller	178	**Understanding calibration and failsafes**	**188**
Mounting a PDB on a drone	180		
Installing the battery	181	Calibrating sensors	188
Configuring the RC transmitter and receiver	183	Failsafe setup and configuration	190
		Setting up a maiden flight	**192**
Telemetry connection	183	Performing a maiden flight	194
Setting up and configuring avionics	**185**	**Summary**	**195**

10

Flight Log Analysis and PIDs   197

Technical requirements	**198**	Interpreting graphs	203
Introduction to flight logs and their applications	**198**	**Understanding PID controllers and their uses**	**211**
Types of logs stored	200	Methods to detect bad PIDs in drones and how to fix them	214
Working with GCS software on logs and their graphs	**201**	Rate controllers in a drone and tuning PIDs	214
Steps to download logs from the flight controller using QGC as the GCS	201	**Summary**	**216**
Analyzing download logs	202		

11

Application-Based Drone Development   217

Technical requirements	**217**	Mission planning for a survey mission	222
Survey-based drones	**218**	Image stitching and post-processing	224
Why aerial surveying is required	218	Final deliverables	225
The role of drones in aerial surveying	219	**Agricultural spraying drones**	**227**
Payloads used in survey missions	219	Spray tank integration with a drone	228
Integrating the payload with the drone	220	Agriculture spraying flight planning	229
The Real-Time Kinematics (RTK) and Post-Processed Kinematics (PPK) positioning systems	222	Precautions and safety	229
		Aerial delivery drones	**230**

Payload release mechanism	230	**System redundancy**	232
Safety regulations in the drone ecosystem	232	**Summary**	233

12

Developing a Custom Survey Drone — 235

Technical requirements	235	**Selecting the drone's components**	237
Geospatial surveys	236	Airframe	238
The use of geospatial surveys in the industry	236	Propulsion system or power trail	239
		Flight stack selection	247
Setting up the requirements for a survey drone	236	GPS module selection	248
		Camera or payload	249
High-quality imaging sensors	236	Telemetry system and selection considerations	250
GPS	236		
Autonomous flight planning	237	**Wiring and assembly**	251
Long flight time	237	**Drone configurations**	253
Real-time monitoring and telemetry	237	Motor and ESC calibration	260
Durability and weather resistance	237	ESC calibration	261
Payload capacity	237	**Summary**	263
All-up weight or gross weight	237		

References — 265

Index — 283

Other Books You May Enjoy — 294

Preface

Hello readers! Welcome to the exciting world of drone development! *Drone Development from Concept to Flight* is a comprehensive guide that takes you on a journey from the initial concepts of drone technology to the exhilarating experience of launching your own **unmanned aerial vehicle (UAV)** into flight. Whether you are a hobbyist, a student, or a professional seeking a deeper understanding of drone technology, this book is designed to be your companion every step of the way.

In the pages that follow, we will delve into various aspects of drone development, covering essential topics such as drone physics, the fundamental components utilized in drone construction, and the intricacies of flight control systems such as ArduPilot and PX4 flight stacks. Understanding the physics behind drone flight is crucial for anyone aiming to design and build their own UAV, and this book provides a solid foundation in this realm.

The core of any successful drone project lies in the selection and integration of components. From motors and propellers to sensors and communication modules, we explore the diverse range of hardware that brings a drone to life. Moreover, we delve into the sophisticated software systems that govern drone behavior, with a particular focus on the widely used ArduPilot and PX4 flight stacks. These systems form the brains of your drone, enabling it to navigate, stabilize, and execute missions with precision.

To empower you further, we guide you through the intricacies of mission planning using popular ground control software such as Mission Planner, AeroGCS, and QGroundControl. These tools play a pivotal role in defining the trajectory, waypoints, and overall mission parameters of your drone, ensuring successful and controlled flights.

Assembling and configuring a drone can be a challenging yet immensely rewarding task. This book provides step-by-step guidance on the assembly process, accompanied by tips for optimal configuration to suit your specific requirements. From connecting the electronic components to tuning the flight controller, we cover the entire process to help you achieve a successful drone build.

The thrill of witnessing your drone take its first flight is unparalleled. We walk you through this momentous occasion, offering insights into troubleshooting common issues and ensuring a smooth and controlled inaugural flight. Whether you are a novice or an experienced enthusiast, the information provided will guide you toward mastering the art of drone piloting.

In the latter part of the book, we explore specialized applications of drones, with a focus on agriculture and surveying. You will discover how drones are revolutionizing these industries, providing valuable data and insights that were once difficult or impossible to obtain.

Embark on this educational and practical journey with us as we demystify the process of developing drones, turning your concepts into reality and propelling you into the exciting realm of autonomous flight. Happy droning!

Who this book is for

This book is for beginner-level drone engineers, robotics engineers, hardware and design engineers, and hobbyists who want to enter the drone industry and enhance their knowledge of the physics, mechanics, avionics, and programming of drones, multicopters, and UAVs. A basic understanding of circuits, assembly, microcontrollers, and electronic instruments such as multimeters and batteries, along with fundamental concepts in physics and mathematics, will be helpful for reading this book.

What this book covers

Chapter 1, *Getting Started with UAV and Drone Engineering*, serves as the foundational chapter in our comprehensive guide. This chapter provides an accessible introduction to the exciting world of unmanned vehicles and drone engineering. The chapter further introduces the key components used in drone construction, paving the way for a hands-on understanding of hardware integration. Whether you are a novice or an enthusiast, this chapter is your gateway to the thrilling journey of developing drones from concept to flight.

Chapter 2, *Understanding Flight Terminologies and Physics of Motion*, unveils the critical elements of drone flight. This chapter decodes the intricate language of aviation, clarifying terms essential for any drone enthusiast. You will delve into the physics of motion, gaining insights into lift, thrust, drag, and gravity—the fundamental forces shaping UAV dynamics. With a focus on clarity, this chapter lays the groundwork for comprehending the intricacies of drone movement and sets the stage for successful flight control system implementation.

Chapter 3, *Learning and Applying Basic Command and Control Interface*, is a pivotal chapter introducing readers to **ground control station** (**GCS**) software and hardware. Delving into the remote controller interface, the chapter provides hands-on guidance for users to master the crucial aspects of controlling roll, pitch, and yaw. This essential knowledge empowers enthusiasts to navigate and manipulate their drones with precision, making it an indispensable resource for beginners and those seeking to enhance their drone piloting skills.

Chapter 4, *Knowing UAV Systems, Sub-Systems, and Components*, is a comprehensive chapter unraveling the intricacies of UAVs. You will explore the diverse landscape of propulsion systems, power systems, and navigation systems, gaining a nuanced understanding of each crucial component. From delving into the mechanics of propulsion to understanding the intricacies of power distribution, this chapter provides essential insights for enthusiasts and professionals alike. Elevate your comprehension of UAV technology with this in-depth exploration of systems, sub-systems, and components.

Chapter 5, *Sensors and IMUs with Their Application in Drones*, is a vital chapter exploring the world of sensors and **inertial measurement units** (**IMUs**) in drone technology. You will uncover the significance of GPS and compass integration, gaining insights into how these sensors contribute to precise navigation and orientation. This chapter serves as a guide for understanding the role of these critical components in enhancing a drone's capabilities, making it an indispensable resource for enthusiasts, engineers, and professionals seeking to optimize their UAVs for accurate and reliable performance.

Chapter 6, Introduction to Drone Firmware and Flight Stacks, unveils the core of drone intelligence. This chapter provides a comprehensive overview of drone firmware, focusing on popular flight stacks such as ArduPilot and PX4. You will gain insights into the intricate software systems that govern navigation, stabilization, and mission execution, laying the foundation for successful drone development. Whether you're a novice or an enthusiast, this chapter is your gateway to understanding the brains behind your drone's operations.

Chapter 7, Introduction to Ground Control Station Software, provides a comprehensive overview of various GCS software, which is essential for effective drone mission planning and execution. You will explore popular tools such as Mission Planner, AeroGCS, and QGroundControl, gaining insights into their functionalities and applications. Through real-world examples, this chapter equips enthusiasts with the knowledge needed to navigate and harness the capabilities of different GCS software, ensuring a solid foundation for successful drone operations. Whether you're a novice or an experienced user, this chapter is your gateway to mastering the tools that facilitate seamless communication and control between the operator and the UAV.

Chapter 8, Understanding Flight Modes and Mission Planning, is an important chapter that dives into the different flight modes essential for drone operation, such as manual, position, and hold modes. You will gain insight into the functionalities of each mode, empowering you to tailor your drone's behavior to specific needs. This chapter not only covers the intricacies of flight modes but also introduces the key principles of mission planning, providing a comprehensive guide for enthusiasts to execute precise and automated UAV missions. Whether you're a beginner or an experienced pilot, this chapter is your gateway to mastering the dynamic capabilities of drone flight.

Chapter 9, Drone Assembly, Configuration, and Tuning, helps guide you through the intricate process of assembling and wiring drone components. From motors and propellers to sensors and flight controllers, the chapter provides step-by-step instructions for seamless integration. You will also delve into the art of configuration, ensuring optimal settings for your specific drone requirements. This comprehensive guide is essential for both beginners and enthusiasts seeking to master the technical intricacies of bringing a drone to life.

Chapter 10, Flight Logs Analysis and PIDs, helps in exploring the post-flight analysis of drone performance through flight logs. You will gain insights into interpreting and utilizing **proportional, integral, and derivative** (**PID**) controllers to fine-tune your drones. This chapter equips enthusiasts with the tools to optimize flight characteristics, making it an indispensable resource for enhancing drone control and performance.

Chapter 11, Application-Based Drone Development, is a focused chapter catering to the diverse needs of survey, agriculture, and delivery drone enthusiasts. You will gain insights into tailoring drone technology to specific applications, from optimizing surveying missions to enhancing agricultural practices and efficient delivery operations. This chapter serves as a valuable resource for harnessing the full potential of drones in various real-world scenarios.

Chapter 12, Development of Custom Survey Drone, is a specialized chapter focused on tailoring drones for surveying applications. You will gain insights into selecting and integrating components specifically suited for surveying tasks, enhancing data collection precision. This chapter serves as a practical guide for enthusiasts and professionals alike, providing the knowledge needed to design and deploy custom survey drones for diverse applications.

To get the most out of this book

Before diving into this book, you should have a basic understanding of physics principles related to drone flight. Familiarity with electronic components and their functions is recommended, along with reading the datasheets, as the book extensively covers component selection and assembly. Additionally, a cursory knowledge of ArduPilot and PX4 flight stacks, mission planning tools such as Git and GitHub, as well as a general grasp of drone assembly, configuration, and flight procedures will enhance your experience. Enthusiasts seeking insight into agriculture and survey drones will find the book more accessible with a foundational understanding of these concepts.

Software/hardware covered in the book	Operating system requirements
AeroGCS	Windows, macOS, or Linux
Mission Planner	Windows
QGroundControl	Windows, macOS, or Linux

If you are using the digital version of this book, we advise you to type the code yourself or access the code from the book's GitHub repository (a link is available in the next section). Doing so will help you avoid any potential errors related to the copying and pasting of code.

It is recommended to read, study, and understand the breadth of knowledge that the book covers. Try developing your own drone after going through this book.

Conventions used

There are a number of text conventions used throughout this book.

`Code in text`: Indicates code words in text, database table names, folder names, filenames, file extensions, pathnames, dummy URLs, user input, and Twitter handles. Here is an example: "Mount the downloaded `WebStorm-10*.dmg` disk image file as another disk in your system."

Bold: Indicates a new term, an important word, or words that you see onscreen. For instance, words in menus or dialog boxes appear in **bold**. Here is an example: "A **Fast Fourier Transform** (**FFT**) graph is used to visualize signals. In log analysis, this is used to convert vibration into amplitudes and display it as a frequency in the frequency domain."

> **Tips or important notes**
> Appear like this.

Get in touch

Feedback from our readers is always welcome.

General feedback: If you have questions about any aspect of this book, email us at `customercare@packtpub.com` and mention the book title in the subject of your message.

Errata: Although we have taken every care to ensure the accuracy of our content, mistakes do happen. If you have found a mistake in this book, we would be grateful if you would report this to us. Please visit `www.packtpub.com/support/errata` and fill in the form.

Piracy: If you come across any illegal copies of our works in any form on the internet, we would be grateful if you would provide us with the location address or website name. Please contact us at `copyright@packt.com` with a link to the material.

If you are interested in becoming an author: If there is a topic that you have expertise in and you are interested in either writing or contributing to a book, please visit `authors.packtpub.com`.

Share your thoughts

Once you've read *Drone Development From Concept to Flight*, we'd love to hear your thoughts! Scan the QR code below to go straight to the Amazon review page for this book and share your feedback.

`https://packt.link/r/1837633002`

Your review is important to us and the tech community and will help us make sure we're delivering excellent quality content.

Download a free PDF copy of this book

Thanks for purchasing this book!

Do you like to read on the go but are unable to carry your print books everywhere?

Is your eBook purchase not compatible with the device of your choice?

Don't worry, now with every Packt book you get a DRM-free PDF version of that book at no cost.

Read anywhere, any place, on any device. Search, copy, and paste code from your favorite technical books directly into your application.

The perks don't stop there, you can get exclusive access to discounts, newsletters, and great free content in your inbox daily

Follow these simple steps to get the benefits:

1. Scan the QR code or visit the link below

https://packt.link/free-ebook/9781837633005

2. Submit your proof of purchase
3. That's it! We'll send your free PDF and other benefits to your email directly

Part 1: Fundamentals of Flight Engineering

This part provides an introduction to unmanned aerial systems, that is, drones, including the different types and the engineering behind them. It also introduces the basic terminology that is used in the field of aerial system development and how it is important in deciding the attitude of the system in 3D air space.

This part has the following chapters:

- *Chapter 1, Getting Started with UAV and Drone Engineering*
- *Chapter 2, Understanding Flight Terminologies and Physics of Motion*
- *Chapter 3, Learning and Applying Basic Command and Control Interface*
- *Chapter 4, Knowing UAV Systems, Sub-Systems, and Components*

1
Getting Started with UAV and Drone Engineering

Unmanned aerial vehicle (**UAV**) systems have become a buzzword globally, as the new technology is revolutionizing every sphere of life from civil to defense and from small photography companies to big industries and security. In this chapter, we will understand the various unmanned systems available with a special focus on UAV systems, their anatomy, and their uses in different domains. By going through this chapter, we will understand the major systems, subsystems, and components of a drone and their function, which will help us understand the system as a whole and also its parts. By the end of this chapter, you will be well versed in the types of drones that are available and their major components with applications, which will give you a push to understand later aspects of the book.

In this chapter, we are going to cover the following main topics, which will help us to bifurcate the chapter and understand it in a better way:

- Introduction to unmanned systems – unmanned ground, air, and water vehicles
- Types of drones and their relevance to applications
- Major mechanical and structural components of a drone
- Avionics systems and subsystems of drones

Introduction to unmanned systems – unmanned ground, air, and water vehicles

As the world is going through innovations and new developments, technology has risen and things have scaled down to a major extent. Any new technology that arises comes from security and defense requirements and is later integrated into the civil world, and unmanned technology is no different.

Manned systems were too big for intelligence gathering and too risky to be used in gathering intelligence from the enemy. Also, a huge price of man and machine was paid if engaged in combat or ambush. To avoid this situation, unmanned technology was born to cater to this purpose on land, air, and sea, now known as **unmanned ground vehicles (UGVs)**, UAVs, and **unmanned water vehicles (UWVs)**, respectively.

Various unmanned vehicles

UAVs have evolved over time and have taken shape, as we are seeing with day-to-day vehicles. As we have cars and trucks for land, submarines and ships for the sea, and airplanes for the air, similarly, we have unmanned systems for all three domains – land, water, and air – described as follows:

- **UWVs**: UWVs are uncrewed submarines that are used to travel deep inside the water. They travel up to long ranges deep inside water and are operated by ground crews from far away. These unmanned submarines are autonomously driven water vehicles that travel on a predefined path at a predefined depth and return after completing missions.

    These are used for underwater surveillance using RGB cameras or doing bathymetric surveys using **Light Detection and Ranging (LiDAR)**. The same is being used in defense for reconnaissance and monitoring sea waters.

- **UGVs**: UGVs are similar to UWVs, with the difference that these vehicles travel on the earth's surface rather than in air and water. These are operated with the crew members sitting far from the base stations. These are operated in manual and autonomous modes on a predefined path, speed, and route and come back to their original place.

- **UAVs**: UAVs are vehicles that fly in the air without onboard crews and with the help of onboard sophisticated sensor systems. These unmanned systems are controlled by ground-based controlled systems crews controlled by long-range antennas. These systems are used for civil and military applications such as crowd monitoring, aerial surveys, and agriculture. We will understand more about these systems in the coming chapters.

History and evolution of drones

The history of UAVs, or drones, is quite interesting. The idea of pilotless flying machines dates back to the early 1900s when humans began imagining the chances of an unmanned flight. In their early days, drones were primarily used for military purposes, such as reconnaissance and surveillance, and they proved valuable in situations where sending human pilots was risky.

As development went through advanced phases, drones became smaller and more sophisticated, and various new applications evolved. The evolution of UAVs rose dramatically in the 2000s with the introduction of consumer drones, whose applications were beyond military use. Suddenly, drones became popular among hobbyists, photographers, and filmmakers, offering a new perspective from the sky.

In recent years, drones have been introduced to various other applications as well, where they are now employed in diverse fields, including agriculture for crop monitoring, search and rescue operations, environmental monitoring, and even delivery services. The history of UAVs reflects a journey from military-focused beginnings to becoming versatile tools with widespread civilian applications, showcasing the remarkable evolution of unmanned aerial technology.

Need for an unmanned system

As the era advances, the demand for different datasets and intelligence is growing. Earlier, due to non-availability of the technology, such things were done with the help of man and machines. Now, as technology is rising, drones can reach where man and machine cannot via the ground with less cost and less effort.

The following are key reasons why UAVs have become a key requirement over manned aircraft:

- **Easy reach**: UAVs have reached remote areas that man or ground vehicles failed to reach easily with less cost and effort.
- **Easy transportation**: Unmanned systems are highly scalable and available in all shapes and sizes for a variety of work as compared to manned aircraft. Due to their extremely small shape and size, they can be easily transportable in a small form factor, which makes them smaller, simpler, and smarter.
- **Less power consumption and easy maintenance**: Being small and rough, these devices consume less power, which makes them more economical.
- **Economical**: Drones prove to be an economical solution for many aerial applications such as surveys and surveillance as compared to other applications.

What are unmanned aerial systems?

An **unmanned aerial system** (**UAS**) is an uncrewed aerial platform being operated by an avionics system over a wireless network by a remote crew. It comes in all weights, sizes, and performances. It comes with different types of vehicles that are used for different types of applications. UAVs are also known as **remotely piloted aircraft systems** (**RPAS**). In the next section, we will understand the various applications for which UASs are used.

A few of the applications of UAVs in current scenarios include the following:

- **Civilian uses of drones**:
  - **Aerial photography**: The use of drones has enabled filmmakers and cinematographers to capture high-quality video from different angles and heights, which was once very difficult and expensive. We can see the use of drones for videography in functions and weddings due to their small size, easy handling, and cost efficiency.

- **Asset inspection**: Earlier, the inspection of huge assets such as windmills, pipelines, power transmission lines, bridges, and the like was difficult as the reach of humans was limited and the execution of tasks was costly and time-consuming. After the evolution of drones and their capacity to carry payloads such as Lidar and cameras, the inspection of these assets has become easy and cost-efficient due to the small size of the drones and their reach to places humans can't. This helps to inspect assets closely and take measures in a timely manner.
- **Wildlife conservation**: Drones help to keep an eye on wildlife spread across a larger area within minutes. They give a real-time video feed to the operator of the landmass and aid monitoring in the area. This saves a lot of time and effort compared to when people have to physically monitor the area. Drones also can issue warnings, take a closer look at areas of wild animals, and report if any critical incident has taken place.
- **Agriculture surveys**: Special multispectral and hyperspectral camera-equipped drones help to take geotagged imagery of crop fields. These images are later used by software to extract the chemical composition of leaves and provide data about the lack of critical minerals in the plants. The data can be used and analyzed to predict crop health at a particular geolocation and take protective measures against it.
- **Aerial surveys**: Drones equipped with high-resolution cameras take geotagged imagery of landmass from the air, and later, these images are used to make high-quality accurate maps to understand the earth's surface. Drones have made this task easy and time efficient due to their small size and require less human effort.
- **Mosquito repellent**: Heavy lift drones are also used to spray mosquito repellent in areas that require efficient mosquito control. Since it's difficult for humans to spray these insecticides evenly across areas where reach is impossible, drones have made this efficient and cost-effective.
- **Cargo drones**: Nowadays, drones are also used in the delivery of goods, which is termed aerial deliveries. Major companies are looking at this as the future of deliveries. Use cases for delivering heavy cargo across remote areas are also being developed.

- **Defense use cases of drones**:
  - **Crowd monitoring**: Drones equipped with speakers and cameras have played a pivotal role in crowd monitoring. A drone gives a live feed of the situation, and a person can instruct the crowd with the help of speakers and investigate the situation with the help of a camera without actually going there.
  - **Surveillance**: Drones equipped with day-night cameras help defense forces keep an eye on critical assets during the day and night. These cameras are equipped with tracking capabilities and lasers to accurately get the geolocation of the target. This helps forces keep a large area under surveillance without any human intervention.

- **Aerial warfare**: As we are seeing in the world, drones have become crucial equipment in warfare. This helps to be more lethal without risking human life.
- **Radio relaying**: Drones are also used as radio signal boosters and repeaters in remote areas where communication is the key tool. Such drones are used in mountain ranges where **line of sight** (**LOS**) is not possible, and a tethered drone is used as a tower for amplifying signals and establishing communications.

The aforementioned are some use cases where drone technology is being used to help reduce the cost and risk of manpower and also reduce the time taken for project completion. We are seeing that different types of drones are being built and used to cater to different application needs. As a fighter plane cannot work as a passenger plane and vice versa, one type of drone does not fit into all applications, hence any system is designed completely from scratch as per the requirements/application/purpose, and so on.

Types of drones and their relevance to applications

As we have gone through use cases that are catered for by the use of drone technology, here, we will study different types of drones that have been built for the sake of different application requirements such as high endurance, long range, high altitude, and so on. By doing this, we will get to know about the different kinds of drones and later build an understanding of their development. A glimpse at various kinds of drones is covered in the following section.

Types of drones and their specifications

In this section, we will study the various types of drones, their key functionality, and how they are different from each other.

Multirotor

A multirotor is a motor-propeller-based drone. The major elements to produce force and lift in the air are motors and propellers attached to them. This produces thrust in the drone and helps to lift the system's load.

Based on the number of motors, these drones are classified as follows:

- **Two motors or bi-copter**: A bi-copter is a multirotor with two rotors placed on each side of the center. These systems are unidirectional systems and capable of lifting less load with low endurance and control. Hence, these systems are not used much:

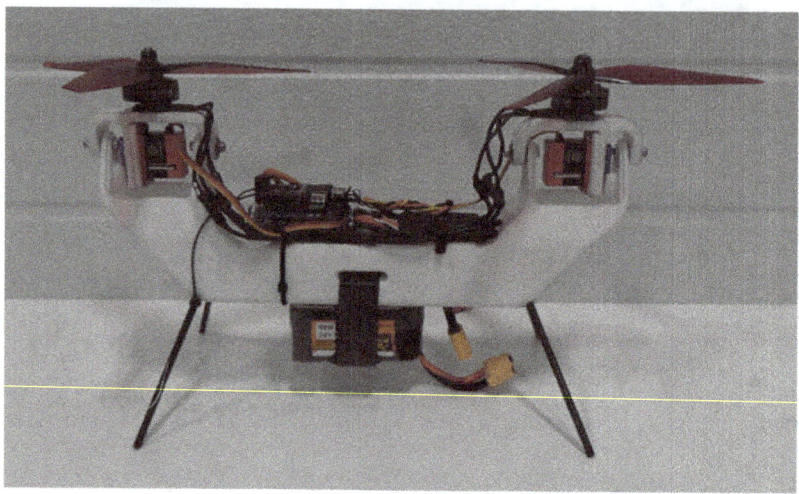

Figure 1.1 – A bi-copter

- **Three motors or tri-copter**: A tri-copter has three arms and lift-generating elements at the end of each arm, placed at 120 degrees to each other. These types of drones are small and not load-carrying but can be used for short distances in inspection and surveillance:

Figure 1.2 – A tri-copter

- **Four motors or quadcopter**: Quadcopters are the most famous drone configurations, used across the world. These are considered the most stable and easy-to-control drones. The dynamics of the drones make them easy to maneuver across the 3D space and give them stability and speed across long ranges and high altitudes. These configurations are also used to carry up to a few kilograms of payload with them:

Figure 1.3 – A quadcopter

- **Six motors or hexacopter**: Hexacopters are one of the most famous configurations after quadcopters. As the name suggests, they have 6 arms placed at 60 degrees to each other. These configurations are used for more stable flight and are able to carry more load. Eventually, they offer less endurance under the same power than a quadcopter:

Figure 1.4 – A hexacopter

- **Eight motors or octocopter**: An octocopter, as the name suggests, has 8 arms placed at 45 degrees to each other. It is an extended version of a hexacopter that can lift more weight and comes in large sizes. These are aerodynamically more stable but also more heavy and power-hungry vehicles:

Figure 1.5 – An octocopter

- **Eight motors (in Quad) or octa-quad**: The octa-quad model is not very famous in the commercial market. It is a good configuration carrying more weight in the small configurations. It's a quadcopter with four arms and two motors placed on the top and bottom of each arm for producing more thrust in less form factor. These types of drones are mainly used to maintain a good size-to-weight ratio:

Figure 1.6 – An octa-quad

- **12 motors (in Hexa) or deca-hexa copter**: The deca-hexa, as the name suggests, has 12 motors on 6 arms placed upside down. These configurations are used for larger drones, such as passenger-carrying drones or heavy cargo drones. These have bigger-sized motors and bigger load capacity and form factor:

Figure 1.7 – A deca-hexa copter

The preceding configuration is used as per the decided payload, load-carrying capacity, and applications.

Fixed-wing drone

A fixed-wing drone, as the name suggests, is a standard airfoil wing-based design where the wing serves as a key lift generator for the system and a single motor (push/pull) helps to cruise in the air. The cruise speed helps to generate adequate lift via wings to travel in the air, and control surfaces work to give direction in the air:

Figure 1.8 – A fixed-wing drone

Fixed-wing VTOL or hybrid drone

Vertical take-off and landing (**VTOL**) aircraft, also called hybrid aircraft, is assisted by four motors to lift off in the air and later transition into a fixed-wing aircraft. This type of drone does not require a long runway unlike fixed-wing drones. It take off and land like a multicopter from a single place and cruises like a fixed-wing aircraft:

Figure 1.9 – A fixed-wing VTOL

Tilt-rotor drone

A tilt-rotor drone comes under the category of fixed-wing VTOL hybrid drones that take off and land like a multirotor and cruise like a fixed-wing drone. The only difference between these drones and fixed-wing VTOL drones is that these drones work on the principle of differential thrust and have common motors for cruise and take-off. The same motors are used as take-off lifter motors, change their angle from 90 degrees to 180 degrees during transition, and are used as cruise motors.

These drones give much more efficiency than a fixed-wing VTOL since the number of motors is reduced and power consumption is also reduced:

Figure 1.10 – A tilt-rotor hybrid drone

Hence, we have seen different types of drones. These drones can be built in different weight and size categories, but to differentiate them based on their weight profile, the **Directorate General of Civil Aviation (DGCA)** in India has classified drones into five categories:

- **Nano**: Any drones less than or equal to 250 grams come under the category of nano drones
- **Micro**: Any drones between 250 grams and 2 kilograms come under the category of micro drones
- **Small**: Any drones between 2 kilograms and 25 kilograms come under the category of small drones
- **Medium**: Drones that are greater than 25 kilograms and less than or equal to 150 kilograms
- **Large**: Drones that are greater than 150 kilograms

We have now studied different types of drones based on their structure and application, but we haven't yet studied what's actually inside the drone and the systems it contains. In the following sections, we will understand the composition of a drone and bifurcate it into different categories.

System composition of a UAV

A UAV has many systems and subsystems that enable it to fly in the air and do missions automatically with safety and precision. These systems are a combination of hardware and software that perform their respective tasks to keep the system under control and stable in the air:

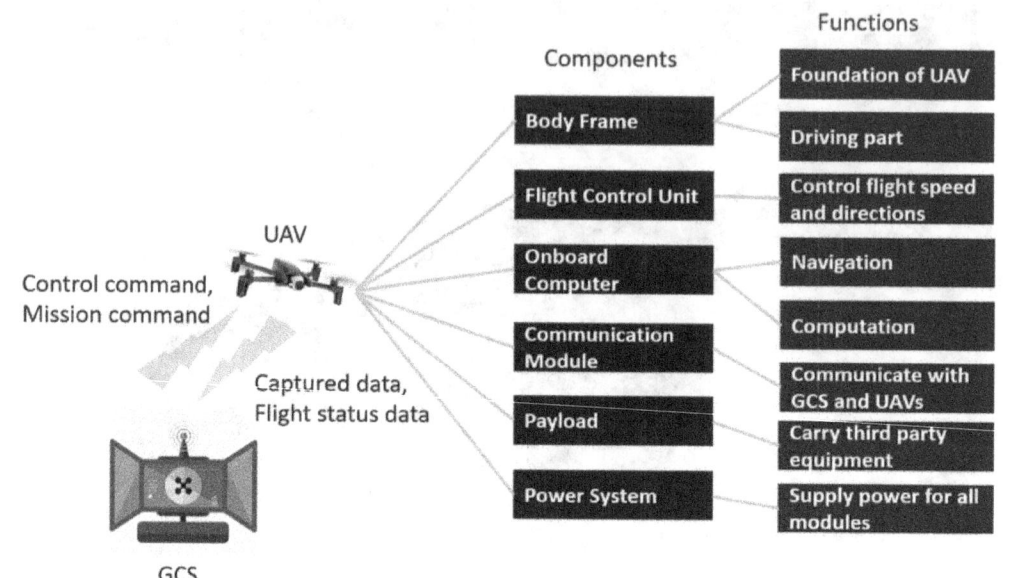

Figure 1.11 – Overview of a drone system

Now that we have seen the system composition of a UAV, let us look at the major mechanical and structural components of drones.

Major mechanical and structural components of a drone

A drone system is a robotic system that is composed of electro-mechanical systems for all its functions. A mechanical system is called a skeleton, the drone under which all the avionics system works. The mechanical system holds the avionics system firmly with it with appropriate strength so that it can take maneuver forces upon it to its limits.

We will study here the major mechanical and structural components of a drone, which are required to hold different parts and have their independent functionalities.

Airframe

The airframe is the main skeleton of a drone, which holds all avionics components in position and helps them to be mounted and fit firmly without any vibrations and loose fitting during the flight. It works as the main body of the drone, which gives the system a proper shape and size, confines all modules, and protects them from direct exposure to the external environment:

Figure 1.12 – A hexacopter carbon fiber airframe

A complete airframe is composed of the following subcomponents:

- **Motor mounts**: Places to hold the motors using screws or other materials:

Figure 1.13 – A motor mount

- **Arms**: Tubes/pipes between the main body and motor mounts are called arms. These are used as a stiff mechanical structure to lift the main body and wiring between motors and the main body:

Figure 1.14 – An arm set

- **Hub**: The hub is the place where the main avionics, such as flight controllers, the **Global Positioning System** (**GPS**), and other components, are placed with due interfacing and connection, which helps the system to get the necessary data to process. Arms are attached to the hub and extended outside:

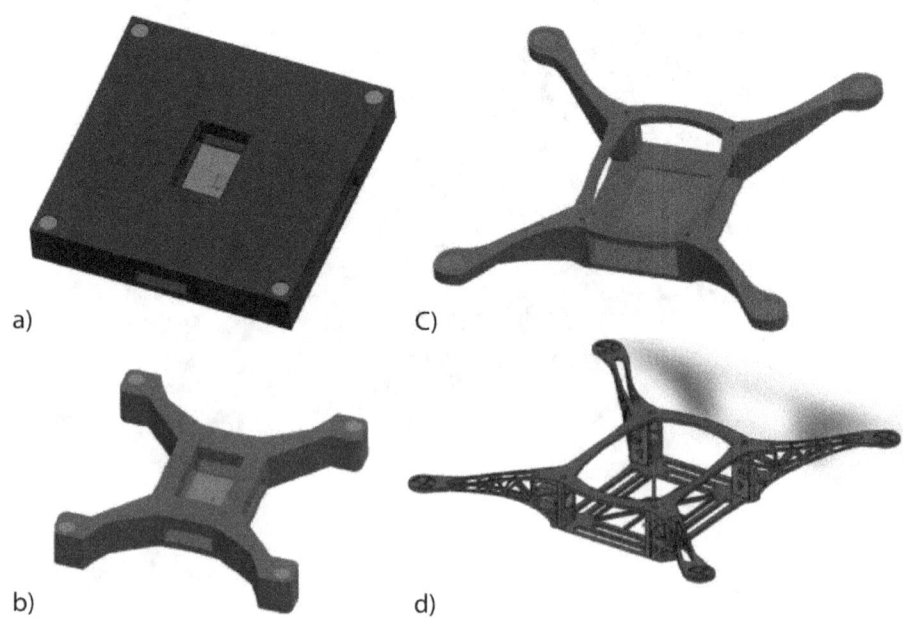

Figure 1.15 – A drone hub

- **Landing gear**: This is also attached to the hub extending downward. This helps the drone to land on different terrains and also keeps adequate ground clearance for the safety of the payload:

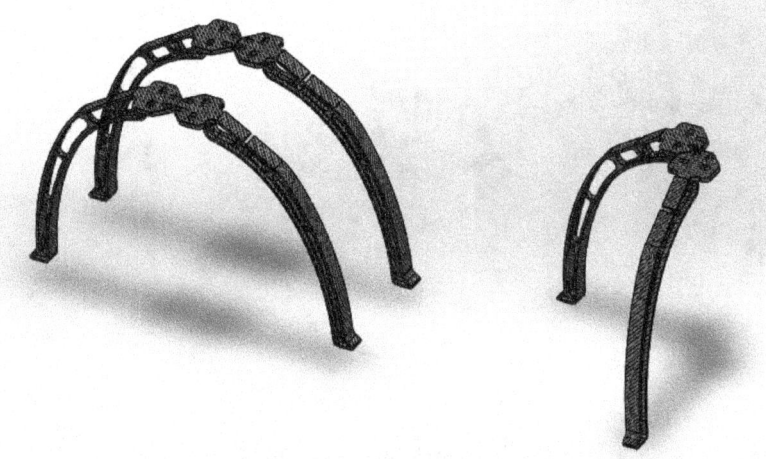

Figure 1.16 – A landing gear

In terms of the features of a mechanical airframe, the following is recommended:

- The airframe should be symmetrical from all aspects on the x, y, and z axes
- Manufacturing of the airframe is to be done from lightweight materials such as carbon fiber, glass fiber, and the like
- A screwless design would be even more helpful for stability and performance

Avionics systems and subsystems of drones

Post the mechanical system comes the avionics system, which is fitted into the mechanical system at appropriate required places to exert thrust and other forces, keeping the center of gravity balanced. These avionics systems include sensors and actuators powered by the power system to exert the necessary force where required and get the necessary task done by the drone:

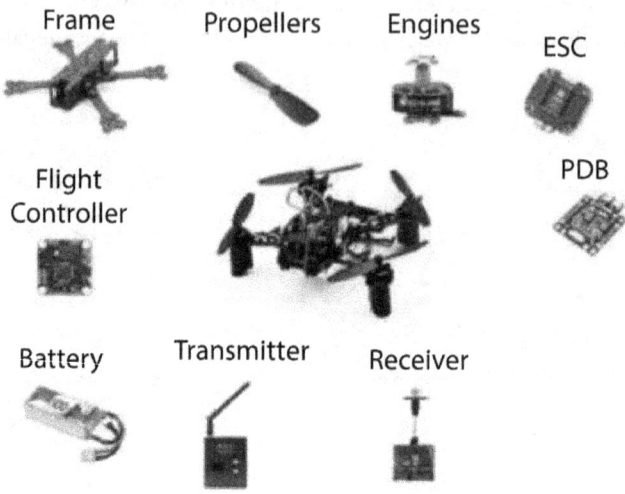

Figure 1.17 – Avionics components

In the following sections, we will have a glance at the major avionics systems and subsystems of a drone that are interrelated.

The propulsion system or drive train of a drone

The propulsion system is responsible for creating the necessary force to lift the system into the air, maintaining the ratios of lift to weight. This is also a reason to maintain stability in the air, fly at high altitudes, and provide speed to the system. This is the most important part and initial building block of drone systems. It also helps in maintaining electrical stability in the system:

Figure 1.18 – A power train

It consists of the following primary parts:

- **Motors**: These are the primary components of the propulsion system. These are **brushless DC (BLDC)** motors, which are far more efficient and lower-power-consumption motors with a long life. These motors rotate at high speed with propellers and produce the necessary thrust to lift the system in the air. Most of the power of the battery is consumed by these motors. These motors are the primary actuators to create motion in the 3D space, travel from one position to another, and lift loads:

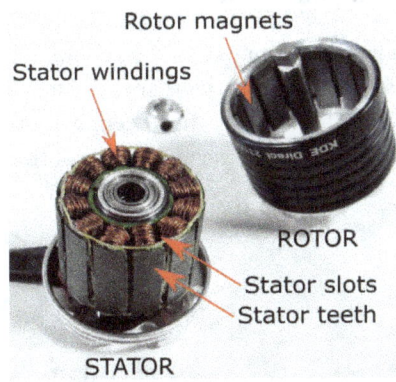

Figure 1.19 – A BLDC motor

- **Propellers**: Propellers are key thrust generators in the multirotor. These are mounted on the motors using screws or other mechanisms. Propellers are manufactured with lightweight materials in such a way that when rotated at high speed, they push the air downward, produce thrust, and lift the system upward. Each propeller produces a set amount of thrust when rotated at a desired RPM, which we will see in the coming chapters:

Figure 1.20 – Propellers

The power system of drones

The power system is the system that is responsible for powering the whole drone with its payload and other peripheral devices. This is the key system that keeps a drone powered, helps in getting the required endurance, and nullifies the effects of high wind or heavy maneuvers. Certain buck-and-boost converters and filters are added to step up and down voltages when required, and current consumption is greater.

These are the major components of the power system of drones:

- **Battery**: The battery is the most important part of the system. It serves as a key source of power delivery to the system. All systems and subsystems are powered by the battery. The battery must be capable of delivering power as required by the system without failure and serve the current delivery. It is also responsible for deciding the endurance of the drone:

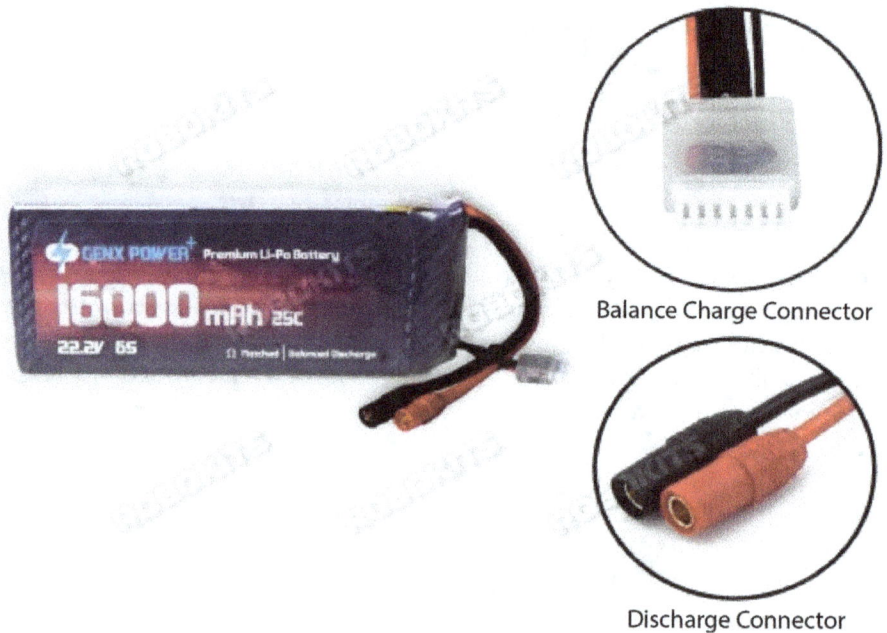

Figure 1.21 – A battery

- **Power distribution board**: The power distribution board, as the name suggests, helps to distribute power to different systems. High-power-consumption elements such as motors take power through this board, and it is evenly distributed among them. A few more filtering elements are added, such as a capacitor to prevent any part from burning due to surges and extra load current:

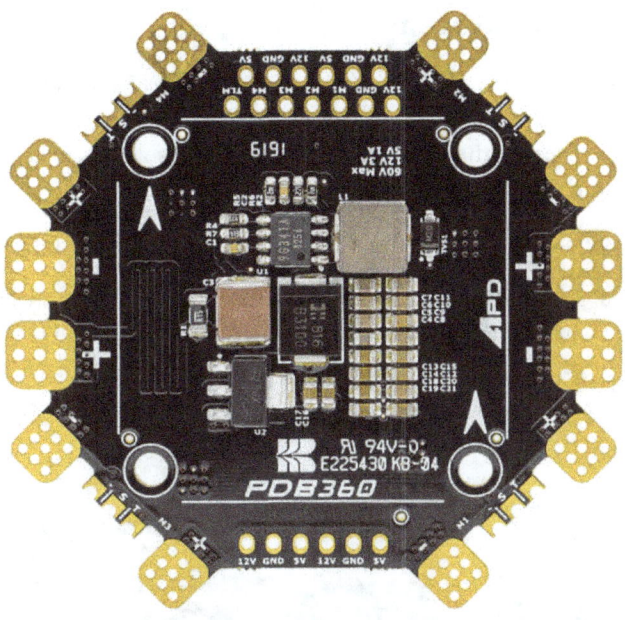

Figure 1.22 – A power distribution board

- **Buck-and-boost converters**: Buck-and-boost converters are used in the circuitry to step up or down the voltage as required by peripheral devices and satisfy the needs of current requirements. These are used as per the power specifications of the payload:

Figure 1.23 – A step-down converter

Command and control system

This is a system that helps connect a drone for command and control and helps to keep the drone in check and in control. It is only due to this system that drones remain in control and can be controlled and operated in the way the pilot wants. This system also helps to keep drone health in check and enables real-time monitoring of sensor data.

The major parts that together make a communication system are the following:

- **Remote control (RC)**: An RC device, or remote controller, is used to fly the drone under the **visual LOS (VLOS)**. It helps to manually fly the system to any place and any orientation. It helps to change the mode of flying and also take control of the system during autonomous missions.

    However, it does not have live monitoring of the system's sensor data and cannot see the live health of the system:

Figure 1.24 – An RC

- **Ground control station (GCS)**: A GCS is an integrated control station that is used for system configuration and ground testing of systems, as well as to keep a check on system health and sensor data, and for command and control over drones for LOS and **beyond VLOS (BVLOS)** applications. A GCS is responsible for mission planning and execution in a completely autonomous mode. It is also responsible for the auto-tuning of systems for smoother flying:

Avionics systems and subsystems of drones   23

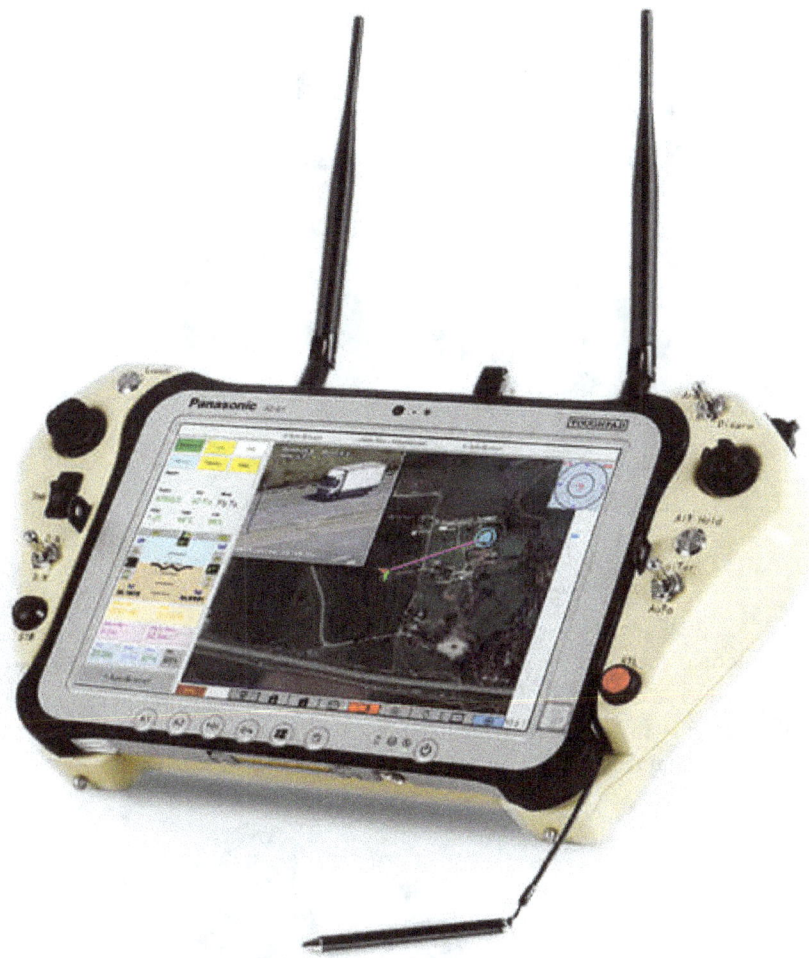

Figure 1.25 – A handheld GCS

- **Radio modem**: A radio modem is a prime device used to communicate between a GCS and a drone. The range of the system depends upon the range at which radio modems communicate. It helps as a key communication link to control and command drones from the GCS. The radio modem helps to provide security to the link and transmit signals on which the GCS and drones communicate:

Figure 1.26 – A radio modem

- **Antennas**: Antennas play a key role in signal transmission and propagation. The antenna gains and orientation decide the range of the system and the degree to which it can move in the air. The type of antenna is an important parameter in deciding the range of a drone or bird, which we will see in later chapters:

Figure 1.27 – An antenna

- **Navigation system**: Post power, propulsion, and communication comes the navigation system. This navigation system is only helpful when driving the system in autonomous mode. The navigation system helps the drone to decide the direction and coordinates and moves in a particular direction autonomously or as guided in the mission.

The drone is able to move in the selected paths and directions with the help of the navigation system, which is supported by built-in sensors. Major components that enable navigation systems are the following:

  - **GPS**: The GPS is a key component of the navigation system. The GPS used in drones has a GNSS such as Galileo, BeiDou, and other similar navigation systems embedded in it, which helps in getting real-time coordinates of the drone to display in the GCS and helps the pilot in getting real-time maps and locations. The same GPS coordinates are used by mission-planning software to decide the autonomous path of the drone and by the drone to navigate to the provided GPS coordinates. The take-off coordinates are also saved into the drone, which helps the drone to do **Return to Launch** (**RTL**) while in failsafe conditions:

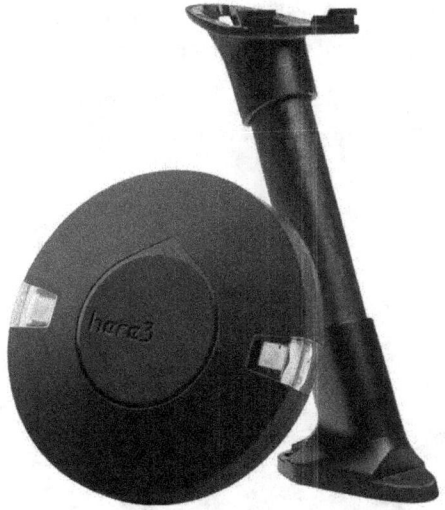

Figure 1.28 – A GPS

  - **Compass or magnetometer**: A compass or magnetometer is a crucial component of the drone navigation system. A GPS gives you the coordinates of the system and the desired go-to location, but a compass gives you the direction in which the drone travels. Drones travel to particular coordinates with the guidance of a compass. The compass gives the heading of the drone where the nose of the drone travels. The compass needs to be calibrated with respect to the magnetic field of the space for proper navigation:

Figure 1.29 – A magnetometer

- **Processing unit or flight controller**: The flight controller is the **central processing unit** (CPU) of the drone. This is called the brain of the drone. The flight controller receives the sensor data from the inbuilt sensors and peripheral sensors and delivers it to the actuators and motors, thus making the drone fly in manual as well as auto mode.

  The flight controller is programmed to behave in a particular manner in various flying modes and missions and under certain conditions, which we will see in the coming chapters:

Figure 1.30 – A flight controller

- **Payload**: A flying machine is to be used for some application and work for which it is designed and developed. The device that executes that particular work while being on the drone is called a payload. For example, a day/night camera is a payload for day/night surveillance (see *Figure 1.31*), a spray tank is a payload for an agriculture spray drone (see *Figure 1.32*), and an RGB sensor is a payload for a survey drone.

The payload is the only object that does the job a drone is intended to do. We will see working with similar payloads later in this book for different applications:

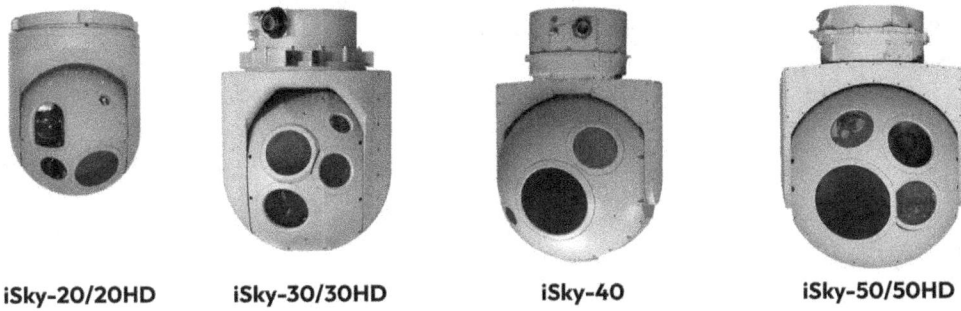

Figure 1.31 – A camera payload

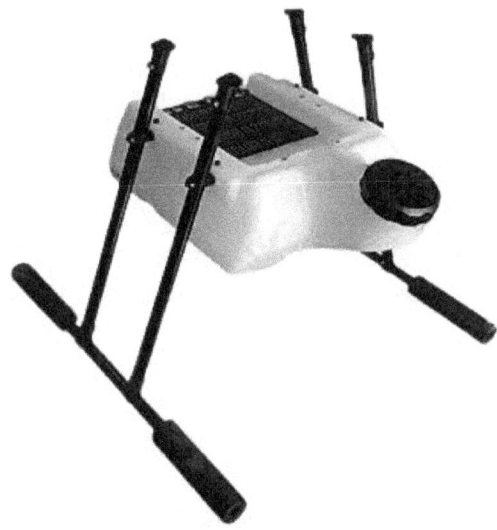

Figure 1.32 – A spray tank as a payload

These are some of the main components that are used in a drone system. In a later chapter, we will look at the specifications used for each of the components and how they can be best optimized to build a drone for a particular specification.

Summary

Having come to the end of this chapter, we have learned about unmanned systems, their development over time, and their requirements. We moved to UAVs and studied their types, important systems and subsystems, and their functionalities.

In the next chapter of this book, we will cover the dynamics of drones, various forces that act on the system when it's in the air, and how to make it safe and balanced. We will also go through various aerodynamic terminologies that will help us to understand things better and look at how we use engineering with the aforementioned components to achieve those forces and build a whole flyable system.

2
Understanding Flight Terminologies and the Physics of Motion

In this chapter, we are going to learn the behavior and stability of a drone in midair. What are the forces that must be acting on the system to maintain the drone's position in the air, and how do we achieve those forces using engineering means? We will also see how the control mechanism of the drone works in the air, with its moment in the free space and major terminologies that are associated with these maneuvers.

Here, we will understand the major forces that act on the body and center of mass across X-, Y- and Z-axes, with major physics acting on the drone. These topics are useful for understanding the basic physics of drones, which is helpful for designers to design an airframe and calculate the load on the hinges of the drone during hover and cruise actions and for avionics development engineers to stay under the load condition provided by the designer. Mechanics and avionics are interrelated concepts that should be optimized for fruitful system development.

In this chapter, we are going to cover the following topics:

- Introduction to thrust, weight, lift, drag, and cruise
- Introduction to the flight axis and its terminologies
- Introduction to the flight physics of a quadcopter
- Engineering to achieve movements across all three axes of rotation

Introduction to thrust, weight, lift, drag, and cruise

Any system that is planned to be in the air is characterized by the physical forces that act upon it in a 3D space, that is, the X-, Y-, and Z-axes. On Earth, the force that pulls an object down is the force of gravity, which has to be overcome first when seeking flight. The force of gravity acts the same towards all objects on Earth, pulling them at an acceleration of 9.8 m/sec^2. When taking this into consideration, all objects have their weight associated with this process, which is acted upon by the force of gravity.

When taking Newton's third law into account, every action has an equal and opposite reaction. The drone experiences a force that is equal and opposite to the force of gravity acting on the body. Let's see the various terminologies associated with this and how we can cater to it.

Thrust

Thrust is the reaction force produced by the rotation of motors and propellers in the drone. This force acts in opposition to the force of gravity in a multi-rotor drone to overcome the weight of the drone. This is an important force that is exerted on the drone to produce lift into the air.

It is measured in Kilograms or Newtons. The thrust produced in the drone is dependent on the size and speed of the rotation of the propellers and the total voltage of the battery used in the drone. The total force applied to the drone has to be double the force applied to the drone by gravity so that it handles the cruise and lift of the drone seamlessly.

Technically, the relationship between the thrust force produced by each propeller to the speed of the motor in RPM or rad/sec is quadratic. In other words, thrust is proportional to the square of the angular speed of rotation.

Weight

The weight of the system is the most important and basic parameter for system design, which helps us to calculate the other parameters, such as thrust, power, and so on. The total weight of the drone includes the weight of the airframe, battery, motors, propellers, and payload. There are different types of weight when referring to drones. They are as follows:

- **All-up weight or gross weight**: This weight is known as the weight of the system taking off into the sky at the moment. This weight varies from payload to payload and flying conditions. This weight would always be less than the **maximum or gross take-off weight** (**MTOW**) of the system to be operated under safe limits.

- **Maximum gross take-off weight**: This weight is the maximum weight of the drone that can be safely lifted into the air without affecting the air-worthiness of the system. The MTOW is dependent on many factors, such as design, structure, and avionics limits. The MTOW is defined by considering the weight of all types of payloads, crossing beyond which the system will not be safe to fly.

Lift

Lift is a vertical force produced by the action of thrust on a multi-rotor drone, which helps the system lift upwards. To produce the necessary amount of lift force for the drone, an adequate propulsion system has to be designed, which we will look at in the coming chapters. In the case of multi-rotor drones, lift is generated by the combined effect of all spinning propellers.

Drag

Drag is a kind of aerodynamic frictional force that is produced by the impact of wind on the body of the system. This force is always opposite to the direction of the motion. This force is generated by every part of the drone. The drone cruise speed must be adequate enough to overcome the force of drag.

All these forces are taken into account when calculating and designing a drone for a particular application.

Cruise

As the term suggests, **cruise** means moving in a particular direction with a particular speed. It's called the traveling phase of the system in midair in a particular direction after take-off and includes gaining the required altitude before it starts to land. The speed at which the system travels is called the **cruise speed**.

Introduction to the flight axis and its terminologies

Until now, we have covered flight terminologies. We can say that a multi-copter hovers when the lift force is equal to the weight of the drone and the cruise speed in the air, which can be termed as cruising in three-dimensional (3D) space. To ease the calculations, we have considered a drone working in 3D space under the major three universal axes, namely, the X-, Y-, and Z-axes. These help us design and program the drone. The balance and uniform distribution of mass and forces across these three axes help us to attain stable flight. Let's look at the movement of the drone across these axes and its terminologies more.

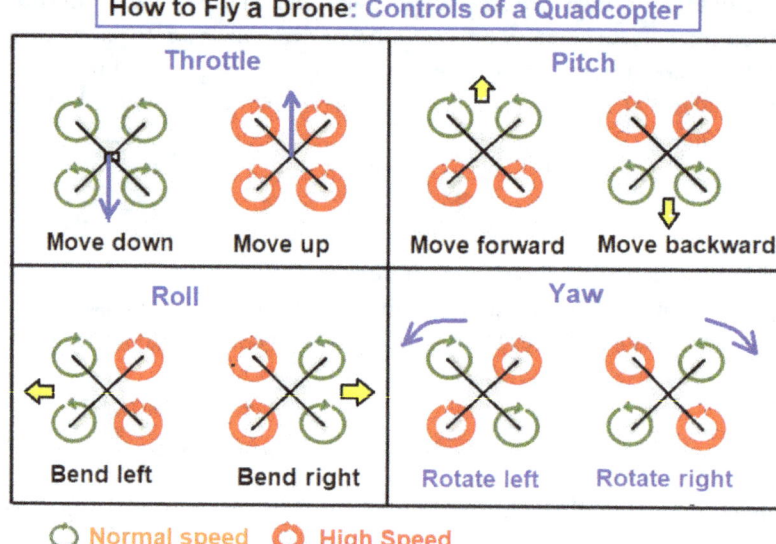

Figure 2.1 – A diagram showing the movement of a drone across different axes

As discussed above, the movement of s drone across the *X*-, *Y*- and *Z*-axes of the 3D plane is coined by **roll**, **pitch**, and **yaw** movement, respectively. A brief description of these movements is given in the following sections.

Roll movement

In 3D space, motion in the *X*-axis is called the **roll movement**. In simple terms, the left and right sideward movements of the system in the space are called roll movements. Roll movements help the system to travel sidewards on the *X*-axis, either left or right.

In multi-rotors, the roll movement is achieved by increasing the rpm of the right motors and lowering the rpm of the left motors identically for a left turn and vice versa for a right turn. In a fixed wing or other planes, roll movement is achieved by the control surfaces called ailerons.

Pitch movement

In 3D space, motion across the *Y*-axis is called **pitch movement**. In simple terms, pitch movement is the forward and backward movement of a drone in 3D space.

In multi-rotor drones, pitch movement is achieved by increasing the RPM of the back two motors and lowering the RPM of the front motors for forward flight and vice versa for backward flight. All these settings and configurations are predefined in the drone firmware, which helps the pilot follow predefined controls.

Yaw movement

Yaw movement is the movement of the drone across the Z-axis in the 3D plane. Yaw movement helps us change the direction of the heading of the drone across any side while in the air. In simple terms, yaw movement helps us change the direction of the heading of the drone at any angle in the entire 360-degree circle. This movement helps us to change the direction of motion in the 3D plane.

Yaw moment in a drone is a function of motor torque, which is applied to the airframe via the rotation of the motors. For clockwise movement in the Z-direction, the anti-clockwise motors rotate at a higher RPM, and for clockwise movement, they rotate at a lower RPM. This helps the drone to rotate clockwise on its axis and vice versa. We will learn about these directions and forces later.

Throttle

Throttle is similar to the throttle we hear about in our day-to-day life regarding bikes or cars. Throttle is the degree of speed at which an object moves. In drones, throttle refers to how fast the motors are rotating, which leads to the faster movement of the propellers, hence generating more thrust and lift. As the thrust increases, the weight of the drone starts to be lifted into the air across the Z-axis, and this stabilizes once the drone has achieved the desired altitude.

For a stable hover, the total thrust should be equal to the total weight of the drone, meaning the drone will hover stably in the air. The altitude of the drone can be increased or decreased by increasing or decreasing the throttle, respectively, to increase or decrease the generation of thrust. For a stable hover, the total thrust should be equal to the weight of the drone at 50% throttle, and for safe aerodynamics, the total thrust capacity of all the motors should be twice the weight of the drone.

The following image further elucidates the movement of a drone in all three axes, shown with reference to the frame:

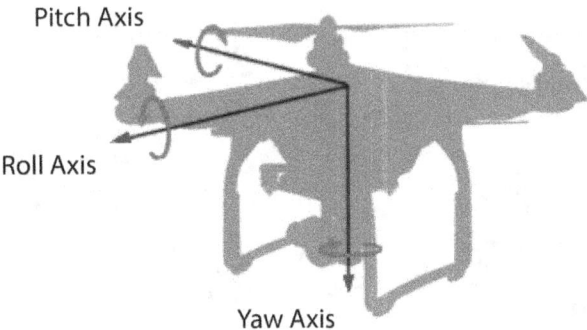

Figure 2.2 – A diagram showing the roll, pitch, and yaw axes movements of a drone

In continuation of the preceding terminologies, there are certain other technical terms that are helpful in drone design.

Thrust-to-weight ratio (TWR)

The thrust-to-weight ratio is a dimensionless quantity and is a performance indicator of the drone. It is the ratio of the total thrust that the system can produce at 100% throttle to the weight of the drone. It is denoted as follows:

$$TWR = \frac{\text{Total thrust produced at 100\% throttle}}{\text{Weight of the drone}}$$

At 50 % throttle, the total thrust produced should be equal to the total weight of the system for a stable hover.

For a safe flight, the effective TWR is 2, which means the total thrust produced should be twice the total weight of the system.

For **first-person view (FPV)** drones or racing drones, the ratio sometimes goes to 3 or higher, as these designs require higher maneuverability and sharper turns.

We will learn more about these technical aspects later in the book when designing the drone from scratch.

Introduction to the flight physics of a quadcopter

For this topic, we will understand a thorough concept of basic flight physics in terms of the hover, roll, pitch, and yaw moments of the quadcopter from scratch. Here, we will understand the forces acting in real time regarding flight and how different forces are acting in different ways; however, the result of all forces decides the movement of the drone in the air.

The following diagram shows the kinematics of a quadcopter and the different forces acting upon it. This helps us to study the different forces acting upon flight:

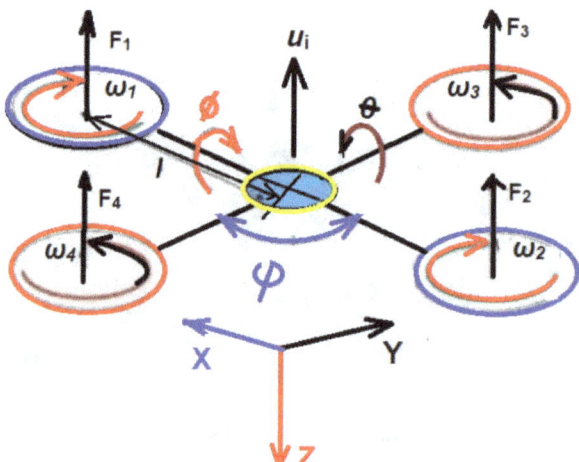

Figure 2.3 – A diagram showing the different forces acting upon the drone

In the preceding diagram, we can see a free-body diagram of a quadcopter, where different forces and motions act on a drone. Let's look at and understand these forces and parameters one by one so we can look forward to working on them.

The weight of the drone enacts a downward force: *M x g (g = acceleration due to gravity)*.

ω x = the speed at which the motors rotate in RPM; hence, ω 1 is the RPM for motor 1, and similarly, ω 2, ω 3, and ω 4 are the speeds of rotations for motors 2, 3, and 4, respectively.

As we can see in the diagram, motors 1 and 2 are operating in the clockwise direction, and motors 3 and 4 are operating in the anti-clockwise direction.

This happens when two diagonal motors (1 and 2) rotate clockwise, and the other two diagonal motors (3 and 4) rotate anti-clockwise; when the motors rotate like this, it produces torque in the opposite direction; hence, two clockwise and two anti-clockwise directions will cancel out each other's torque at the same RPM, making the system stable in the air. Moreover, the same anti-torque is used to give yaw movement (rotation across the Z-axis) to the drone in the air.

Fx is the force produced by each propeller by the rotation of the motor in the upward direction against the force of gravity. Hence *F1*, *F2*, *F3*, and *F4* are the forces produced by motors 1, 2, 3, and 4, respectively.

Ui is the total force produced by the system acting on the center of mass and the center of gravity of the drone by forces *F1*, *F2*, *F3*, *and F4*. Hence, *Ui = F1 + F2 + F3 + F4*.

L is the distance between each motor and the center of mass. *L* is defined to calculate the force acting on the center of mass by each end, which is called the **moment**.

This means that moment M1 by Motor 1 on the center of mass would be *M1 = F1 X L1*.

Similarly, *Mx = Fx X Lx*.

Hence, for an anomaly-free stable flight, the moment for all four motors should be equal and opposite, which cancels out each other, making the resultant moment across the center of gravity zero.

Theta ($\emptyset$) = the angle at which a drone moves in the X-axis, i.e., roll; hence, this is called the roll angle.

Phi (ϕ) = the angle at which a drone moves in the Y-axis, i.e., pitch; hence, this is called the pitch angle.

Sai (ψ) = the angle at which a drone moves in the Z-axis, i.e., yaw; hence, this is called the yaw angle.

Now, the total force, F, produced by each rotor = *Kf X Wx*, where *kf* is proportionally constant, and W2 is the speed of rotation of the motors.

Similarly, the drag moment, *M*, in the rotation of propeller = *Km X Wx*.

Both the forces can be seen in the graph plotted below between the thrust, drag, and RPM.

The following is a graph that depicts the relation between the force produced vs. rotor speed. It also reflects the total speed and moment generated by rotors on the system.

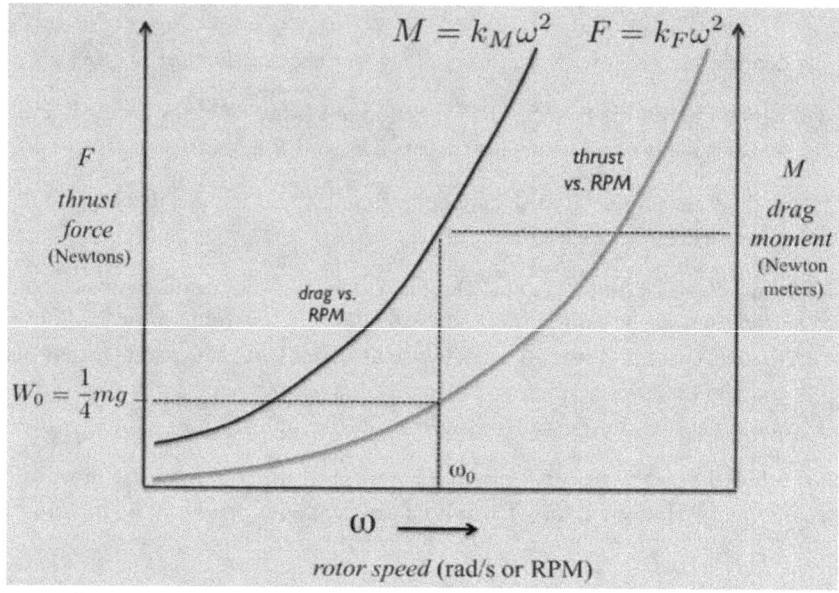

Figure 2.4 – A graph showing the relation between the force and rotor speed

We can say that one-fourth of the total force produced by the rotor is consumed in overcoming the drag caused by the motors; hence, the total effective force produced by the rotor would be $F\_effective = ¼$ (the total thrust produced).

Now, the total resultant force produced by the thrust is the following:

- The hover condition is

    F Total = F1 + F2 + F3 + F4 − mg, which is the resultant force, where $Fi = Kf \times \omega i^2$.

- The resultant moments across the center of mass is

    M = L1 X F1 + L2 X F2 + L3 X F3 + L4 X F4 + M1 + M2 + M3 + M4

    Mx = (F3 − F4) X L; that is, the moment across the x-axis is 0, as the forces would cancel out each other, making the resultant force zero.

    My = (F1 − F2) X L; that is, the moment across the y-axis is 0, as the forces would cancel out each other, making the resultant force zero.

    Hence, all moments acting on the drone are also zero.

So, finally, from these equations, all net resultant forces are zero on the drone, which leads to stable hover flight.

Now, the second law of Newton States that for linear motion, $F = Ma$, i.e., force = the mass of the body multiplied by acceleration.

For rotational motion torque, $T = L \times a$, that is, torque = inertia multiplied by angular acceleration.

Engineering to achieve movements across all three axes of rotation

In the preceding sections, we have studied the physics behind the different motions of the drone, meaning hover and motion across roll, pitch, and yaw. What forces derive these movements and help us to achieve motion across all directions? In this topic, we will learn about the engineering that is applied to the drone to achieve these forces across this direction and how we play around with the mechanics of the drone using this engineering.

Hover motion

The following conditions are to be met for this motion:

- **Weight**: the motors are rotated up to a speed where the total force produced by the rotors becomes equal to the weight of the drone; hence, the resultant force becomes zero, and the drone hovers at a particular place.
- **Moment**: Since the length of all arms is the same and the forces produced by the rotors are also equal and opposite, which cancels out each other, this makes the resultant force zero.

  Additionally, two diagonal motors rotate clockwise and the other two diagonal motors rotate anti-clockwise, which, again, cancels out the anti-torques of each, making the resultant force zero across the Z-axis, and the system stable across the X-, Y-, and Z-axis.
- **Accent motion**: As the throttle is given, the rotors start spinning faster with equal RPM, producing equal thrusts that lead to an increase in the lift of the drone. For an upward-raising motion, the total force produced by the rotors should be more than the weight of the drone; that is, the more force there is, the faster the drone will rise. As the upward force becomes slightly more than the weight of the drone, the drone starts to gain altitude. Now, from the equation of linear motion, the condition for a stable hover flight in equilibrium is $F = Mg$. If F is increased to more than Mg, then there is an acceleration in the upward direction. So, for upward motion, $F > Mg$, that is, $F1 + F2 + F3 + F4 >$ *the weight of the drone*. The faster the forces increase, the faster the acceleration increases, and the drone will keep on gaining altitude until the time the throttle is reduced.
- **Decent motion**: As the throttle is lowered, the rotors decrease in rotational speed, and so the total thrust force also decreases, bringing it to less than the total weight of the drone. Hence, the drone moves downwards. To represent this technically, the weight of the drone should be more than the total force produced by the rotors. That is, $F1 + F2 + F3 + F4 < Mg$.

As the thrust force is slightly less, the drone starts to decrease in altitude and starts moving downwards.

Horizontal movements

The horizontal movements of the drone require some special engineering to play with the forces to provide forward and sideward movement. The following conditions are to be met for this movement:

- **Roll motion**: For a roll motion, the drone tilts to a theta angle on the *X*-axis; thus, the force, *F*, that makes the horizontal and vertical vector components for the motion (that is, F_cos(∅) and F_sin(∅), respectively) that we can see as the force, F, pointing upwards in hover condition. In roll, the direction of *F* is tilted to a theta angle on the *X*-axis, meaning the following:

    Horizontal component = F_cos(∅)

    Vertical Component = F_sin(∅)

    So, the force due to which motion would occur in the x-axis will be given as *F_X_axis = F_cos_theta* since *F = Ma* and *a = F/m* acceleration would be equal to the total force on the drone divided by the mass of the drone. The weight of the drone is to be countered by the vertical component called the F_sin(∅). Hence, we get a roll movement due to these forces.

- **Left side movement**: The two motors on the left side are slowed down, and the two motors on the right side rotate faster equally. This makes the total thrust and the total lift the same without impacting it, but the moment across the center of gravity is increased towards the left side, hence pushing the system towards the left. The speed of the sideward moments would be equal to *F_Cos_theta* towards the left:

    Moment due to left motors < moment due to right motors.

    F_left X L < F_right X L.

- **Right side movement**: The two motors on the right side are slowed down, and the two motors on the left side rotate faster equally. This makes the total thrust and the total lift the same without impacting it, but the moment across the center of gravity is increased towards the right side, hence pushing the system towards the right. The speed of the sideward moments would be equal to *F_Cos_theta* towards the right.

    Moment due to right motors < moment due to left motors.

    F_right X L < F_left X L.

    Therefore, the force on the left side is more than the force on the right side, thus pushing the system towards the right.

Pitch movement or forward-backward movement

The following conditions are to be met for these movements:

- **Forward movement**: The two motors on the front side are slowed down, and the two motors on the back side rotate faster equally. This makes the total thrust and the total lift the same without impacting it, but the moment across the center of gravity is increased towards the front side, hence pushing the system towards the front. The speed of the forward moments is equal to *F_Cos_theta* in forward direction.

   Moment due to front motors < moment due to back motors.

   F_front X L < F_back X L.

   Hence, the force on the back side rotors is more than the force on the front side rotors, thus pushing the system forward.

- **Backward movement**: The two motors on the back side are slowed down, and the two motors on the front side rotate faster equally. This makes the total thrust and the total lift the same without impacting it, but the moment across the center of gravity is increased towards the back side, hence pushing the system towards the back. The speed of the forward moments is equal to *F_Cos_theta* in the backward direction:

   Moment due to back motors < moment due to front motors.

   F_back X L < F_front X L.

   Hence, the force on the back side rotors is more than the force on the front side rotors, thus pushing the system forward.

- **Yaw motion**: The yaw motion is caused by the moment across the Z-axis due to the torque produced by the rotation of the motors. It is both clockwise and anti-clockwise. According to the equation of rotational motion, the moment across the CG is $M = L1 \times F1 + L2 \times F2 + L3 \times F3 + L4 \times F4 + M1 + M2 + M3 + M4 = 0$. At a stable hover, all forces are zero.

   In yaw condition, since *F1, F2, F3,* and *F4* are pointing upwards, it does not have any impact in the moments across the Z-axis.

   So, $M = M1 + (-M2) + (-M3) + M4$, the minus (-) sign indicates rotation in the opposite direction, as two diagonal motors rotate in the clockwise direction and the other two diagonal motors rotate in the anti-clockwise direction.

   For **clockwise yaw movement**, we slow down the two diagonally opposite clockwise motors and speed up the other two diagonally opposite anti-clockwise motors in the same proportion. By doing so, the total thrust remains the same, but the reaction force of the two anti-clockwise motors (in the clockwise direction) is more than the reaction forces of the clockwise motors (in the anti-clockwise direction). Hence, the resultant force is in the clockwise direction, and the airframe rotates in the clockwise direction on the Z-axis.

The rate at which it rotates in either direction is the rate at which the motor speeds are changed.

For **anti-clockwise yaw movement**, we slow down the two diagonally opposite anti-clockwise motors and speed up the other two diagonally opposite clockwise motors by the same proportion. By doing so, the total thrust remains the same, but the reaction force of the two clockwise motors (in the anti-clockwise direction) is more than the reaction forces of the anti-clockwise motors (in the anti-clockwise direction). Hence, the resultant force is in the anti-clockwise direction, and the airframe rotates in the anti-clockwise direction in the Z-axis.

Summary

In this chapter, we have studied the technical terminologies of the aerodynamics and forces that are helpful in steady flight. We have also studied the physics behind the working of the quadcopter and also derived the necessary formulas for the same. By learning this, you can design your own quad or hexacopter. We have seen the necessary engineering behind the quadcopter to achieve the necessary forces to drive it. We are now well-versed in working with and understanding the quadcopter.

In the next chapter, we will study each component of the copter more thoroughly, understanding what the key specifications of the components are, why they are used, how to conceptualize a system, and what the other parts of the system are.

3
Learning and Applying Basic Command and Control Interfaces

In the previous chapters, we learned about various kinds of unmanned systems and their applications. We also studied the basic physics behind drones and how different forces are exerted on the drone through the use of motors and actuators. We also covered the various terminologies, such as roll, pitch, and yaw, and how to achieve those using the forces produced by the actuator.

In this chapter, we'll learn how to command and control a drone so that it can perform a particular task or mission and how to use one in different applications. Moreover, we'll learn about the command-and-control software and hardware and the various types that are used for this purpose. We'll also learn about the basic communication protocols that are used in the industry as communication interfaces between the drone and the **Ground Control Station (GCS)**.

By going through this chapter, you will understand the GCS and the **remote control (RC)** that's used to control the drone and how they work. We'll cover the following topics in this chapter:

- Introduction to GCS hardware and software
- Introduction to different kinds of GCSs
- Introduction to MAVLink and serial protocols
- Introduction to RCs and the different available options
- Introduction to RC protocols

By going through these topics, you'll gain in-depth knowledge of various communication protocols and their usage regarding drones. This will also help you understand the connection that's made between air units (drones) and ground units (GCS) through communication devices.

Introduction to GCS hardware and software

A GCS is a ground-based integrated set of hardware and software that's used to configure the drone and monitor system health, headings, and other parameters by the drone operator. The same GCS is also used to program autonomous missions for the drone and monitor real-time parameters such as height, speed, and direction. It also does a pre-flight check to calibrate its sensors before flight.

The following are the components of a GCS:

- GCS hardware
- GCS software

Let's take a closer look at each.

GCS hardware

GCS hardware is a set of all hardware that's required to run the GCS software and communicate with the drone – that is, the ground data terminal. This GCS hardware is a combination of a laptop/tablet with a human-machine interface, telemetry antennas, interfaces, and video capture cards. GCS hardware is a compatible platform under which all the required software that's used for running the drone system works. This includes the GCS software, telemetry data, and video. A video data terminal or a communication module may or may not be part of the GCS hardware. Some manufacturers provide them together as integrated modules, while others give the user the freedom to choose their own video telemetry and communication modules.

The following are the components of GCS hardware:

- GCS computer
- Multiple screens (optional) for maps, video, mission planning, and monitoring
- Ground data terminal
- Interfacing buttons, switches, and potentiometers
- Battery powering

GCS software

GCS software is a kind of GUI-based virtual cockpit for drone pilots that runs on compatible hardware and communicates with the drone using the **ground data terminal (GDT)**. This helps with planning/editing missions on which the drone operates. It configures modules such as the battery sensors, calibrates the **Electronic Speed Controllers (ESCs)** and RC, and tweaks the performance parameters of the drone.

The following are the major components/windows of a GCS software:

- **Vehicle setup window**

    This includes the settings related to the drone vehicle. It includes airframe setup, sensor calibrations, ESC calibrations, power setup, RC setup, and more. In a nutshell, this window governs all parameters related to system performance and flight.

- **Mission planning window**

    This window includes the maps of the area in which the drone currently is. This is fetched from the internet and reflects the current position of the drone. The type of mission that the pilot intends to fly is planned in this window.

- **Control and command window**

    This window provides various options/buttons to pass commands to the drone, such as takeoff, land, mode change, and return. This helps control the behavior of the drone.

- **Compass and level horizon windows**

    These windows assist the drone in intercepting the heading and level of the drone when it's in the air. The compass window helps the pilot observe the heading of the drone and the direction in which it's moving, whereas the level horizon window helps assert the level concerning the ground and its movement in terms of roll, pitch, and yaw.

- **Video panel**

    The video panel shows the camera feed from the drone directly in the GCS. This helps the pilot observe and extract necessary information from the video.

- **Pre-flight checks window**

    This window helps the pilot check the basics of the drone before flight, such as propellers and motor directions, battery, modes, RTL, and Geofence, if all is set.

- **Log downloads and analysis**

    Most GCS software has a window where the flight logs can be downloaded and used for analysis.

Types of GCS

Based on the application and drones, the GCS hardware can be differentiated into many categories based on their size, form factor, and functionality available. Some of them are as follows:

- **Large vehicle-mounted GCS**: A large vehicle-mounted GCS is a kind of virtual cockpit for drone pilots. It contains all the GCS systems, such as multiple screens for maps and videos. It has an aircraft-like view mounted under a movable vehicle with satellite and microwave antennas for communication with long-range drones. These kinds of GCSs are used in large drones with operational ranges of up to 300 km. These GCSs are high-grade professional systems:

44 Learning and Applying Basic Command and Control Interfaces

Figure 3.1 – A vehicle-mounted GCS system

- **Non-vehicle-mounted portable GCS**: These kinds of GCSs are between a handheld GCS and a vehicle-mounted GCS. These GCSs also have multiple screens and are interfaced with an aircraft-level joystick, but they are man-portable and do not require a vehicle for portability. They are also battery-powered and can be interfaced with multiple UAVs at once. These GCSs do not have a fixed setup but can be managed as per the application's requirements:

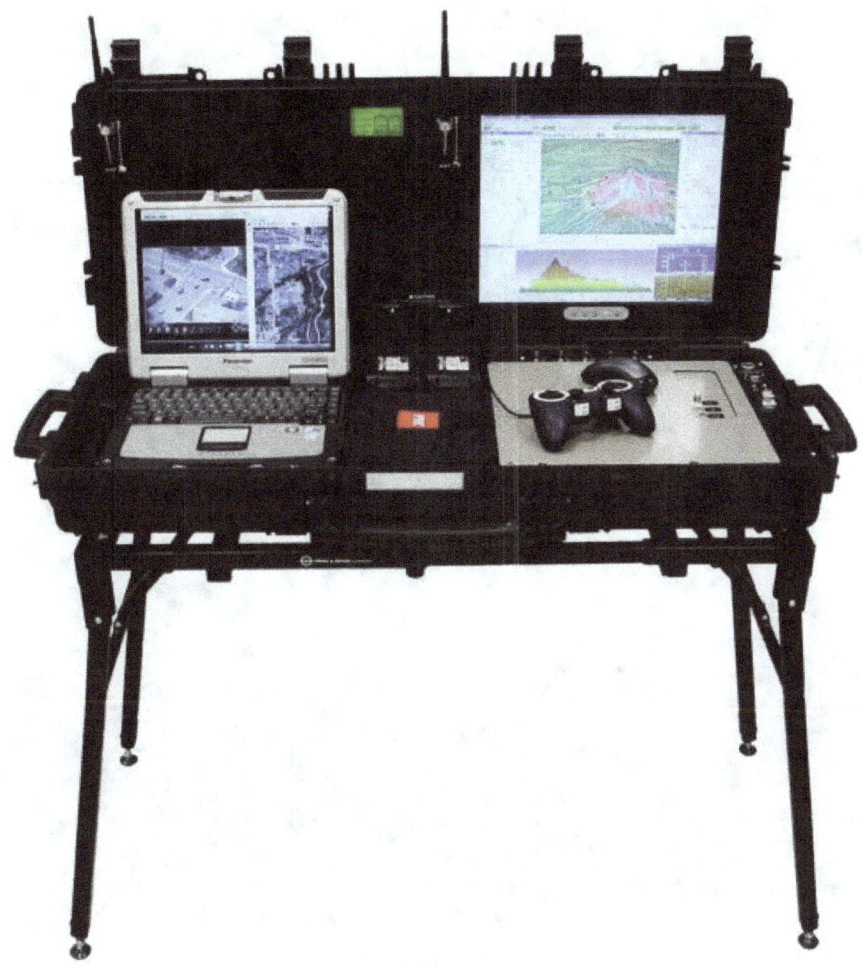

Figure 3.2 – A non-vehicle-mounted GCS system

- **Handheld GCS**: As the name suggests, a handheld GCS is a simple and single entity that's integrated with a built-in radio modem and antenna with a single screen and handheld joystick. These GCSs offer single-handed portability and are used for small and medium-sized drones:

46 | Learning and Applying Basic Command and Control Interfaces

Figure 3.3 – A handheld GCS system

GCS software is something that most companies develop so that it's only compatible with their proprietary systems. This GCS software is not accessible by any other users and is subscription-based. Certain companies develop GCS software that's compatible with multiple kinds of drones. These companies are not drone manufacturers but enrich their GCSs with the required features.

The third category is open source GCS software. This software is developed by the open source community and can be downloaded and used for free by users. The source code of this GCS software is freely available. The users can download the relevant source code and use it to develop features.

Some GCSs are based on Windows, while others are based on Linux systems; some work on both. They are also classified into laptop and mobile-based GCS software. Let's look at some types of GCS software that are classified as desktop or mobile-based. They can be downloaded from the respective websites.

Desktop-based GCS software:

- Universal GCS
- SmartAP GCS
- UrbanMatrix LaunchPad
- Orbit GCS

Mobile/tablet-based GCS software:

- Tower
- Mav Pilot
- Side Pilot

Introduction to MAVLink and other open source telemetry protocols

A telemetry protocol is a wireless serial protocol that enables wireless communication between the drone and the GCS. These protocols are lightweight and require low energy to operate. They are enabled over wireless media using transmitters and receivers.

Various telemetry protocols are widely used in unmanned systems to enable telemetry communication. Here are some examples:

- MAVLink
- ZigBee
- LoRaWAN
- Message Queue Telemetry Transport (MQTT)
- Open-LTE

Let's study these protocols in depth to understand their usage and operation:

- **MAVLink**: MAVLink, short for Micro Air Vehicle Link, is a communication protocol that's used by drones to communicate between various components, such as pilots, GCSs, and other onboard equipment. This is a simple and efficient way for various parts of the drone to communicate with each other and share information about the drone's status, sensor data, and operational instructions. MAVLink ensures the integration and communication between the different components of the drone system.

- **ZigBee**: Zigbee is an advanced Bluetooth-based serial wireless protocol with low energy requirements that operates at 250 Kbps at 2.4 GHz, with a range of 10 to 100 m. Zigbee has anti-interference and high-reliability features. Zigbee protocols have a 128-bit encryption standard algorithm and message integrity code.
- **LoRaWan**: LoRaWan is a long-range telemetry protocol with a low data transmission range from 0.3 Kbps to 50 Kbps that works on the minimal possible infrastructure, with a low cost of operation. It uses an AES algorithm for encryption and authentication.
- **MQTT**: MQTT is a short-range protocol that uses a publish and subscribe model with high reliability, which makes it lightweight and efficient. This protocol works on mesh topology.
- **Open-LTE**: LTE stands for **Long-Term Evolution**. It's based on cellular network technology, which gives it high reliability and satellite coverage. It also supports hard QoS with end-to-end QoS and a guaranteed bit rate for radio bearers. Its data transmission ranges from 20 Mbps to 100 Mbps.

Introduction to RCs and the different available options

A drone system can be controlled simultaneously by multiple channels, such as radio frequency, LTE, and satellite radio, on different protocols. Here, each channel has equal control over the drone and follows the command and control of each system.

Similarly, an RC works on the **serial bus** (**SBUS**) protocol simultaneously with the MAVLink protocol, which runs on serial protocols. Operation through RC gives you a second channel for controlling the drone.

RC is a handheld device that works on radio frequency signals, which are used to control the drone's behavior. It helps it take off, land, fly manually, and have its payload controlled from a safe distance.

This means that even if the GCS control fails, this is an alternate channel by which a pilot can take manual control of the drone and send basic commands to the drone, such as return and land. During long-range operations, the RC connection gets disconnected, but the long-range GCS (MAVLink) connection remains and can be used for command and control. The handheld GCS system transmits the RC data over MAVLink, which helps with gaining long-range RC control of the drone.

The prime need for using an RC transmitter is to take manual control of the drone, where a pilot flies a drone rather than using autopilot mode. A pilot has complete freedom to govern and fly the drone rather than do so in a predefined mission mode in case they need to perform an emergency recovery. Hence, for hobby flying, GCS is not required. Hobby flying can be done with a simple RC that flies the drone manually. Some modes can be configured so that the RC sends important commands such as return, land, and so on.

Components of the RC transmitter system

There are two main components of the RC transmitter system:

- RC transmitter
- RC receiver

Let's look at them in more detail.

The RC transmitter (ground unit) is also simply called RC. A drone transmitter is a handheld device that helps transmit the real-time commands of the pilot to the drone. It helps them take off, land, and fly the drone manually and move and control the payload. The following figure shows a FlySky RC, which is used as a transmitter for a drone; it has a joystick, potentiometers, and switches that can be configured to various channels so that various functions of the drone and payload can be controlled. It also has a screen that displays valuable information:

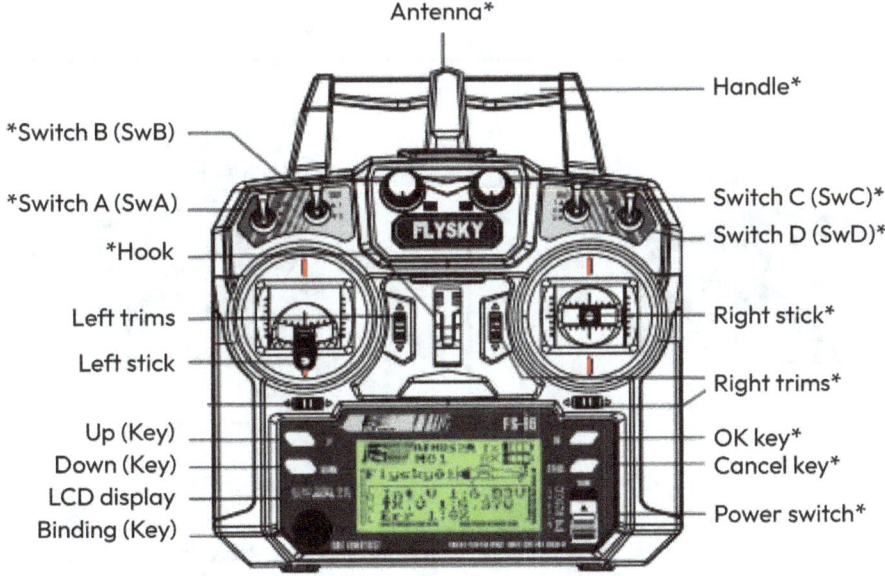

Figure 3.4 – A FlySky RC

The RC receiver (air unit) is the antenna; it receives the signals that are transmitted from the transmitter, decodes them, and sends them to the respective channels for control and operations.

Once the flight controller receives the signals from the RC, it guides them to the peripheral devices connected to them, thus making everything work. The typical range of an RC is 1 km to 1.5 km.

Modes of working of an RC controller

The throttle, roll, pitch, and yaw of the drone are controlled by the joystick on the RC transmitter. The mode of control is governed by the assignment of roll, pitch, yaw, and throttle. Various modes of working are shown in the following figure:

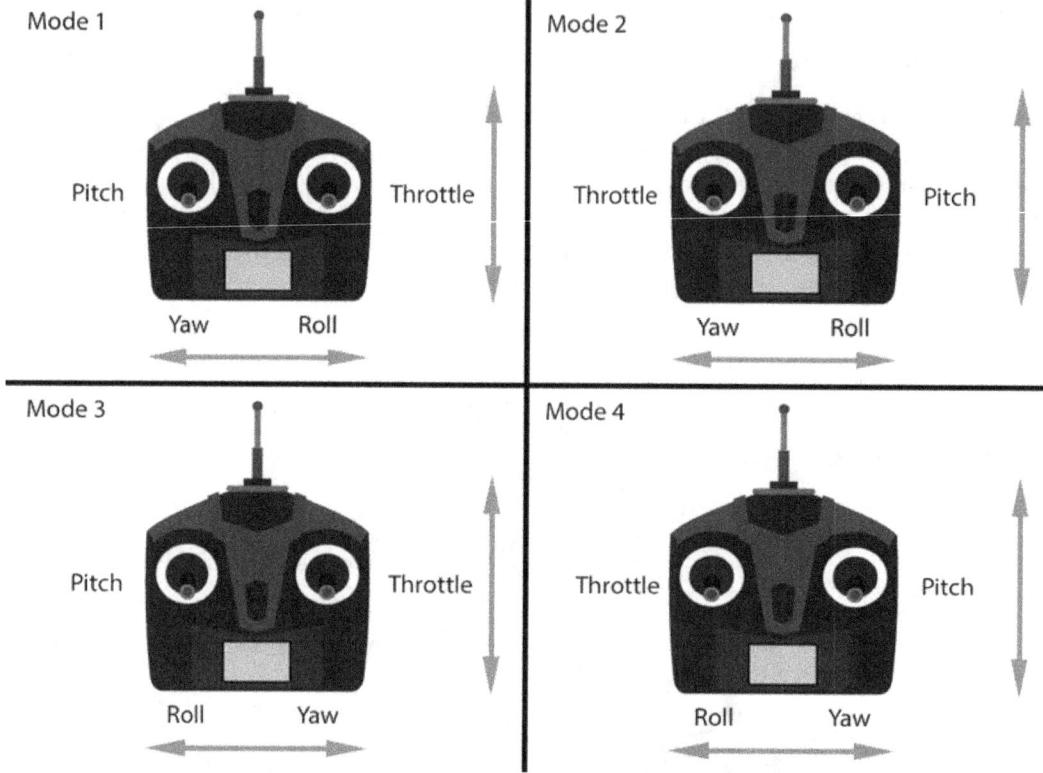

Figure 3.5 – The different modes of working of a transmitter

As you can see, in different modes, different joystick movements are used. In most cases, we would use mode 2 on the transmitter side.

The following figure provides a brief description of the channels that are associated with each stick when using RC mode 2:

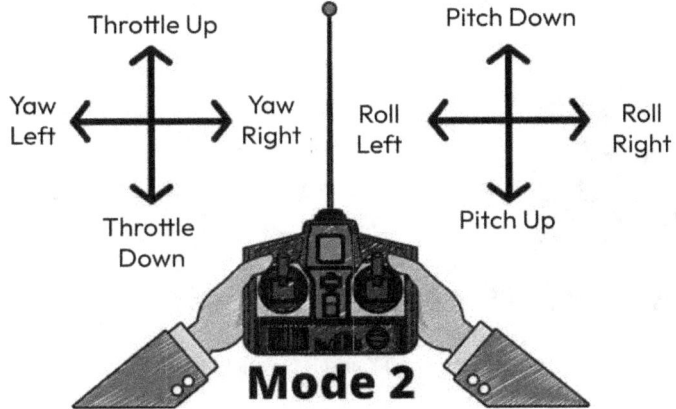

Figure 3.6 – The different channels of each stick when using RC mode 2

Here, we have the following:

- **Roll**: The left and right-hand side movements of the drone are governed by the right-hand stick on the transmitter.
- **Pitch**: The forward and backward movements of the drone are governed by the up and down movements of the right-hand stick on the transmitter. This is known as pitch movement.
- **Throttle**: The upward and downward movements of the drone are governed by the top and bottom movements of the left-hand stick of the transmitter. This is called throttle and is used for height adjustment.
- **Yaw**: The clockwise and anti-clockwise movements of the drone are governed by moving the left stick of the transmitter left and right.

Major RC protocols

RC protocols can be divided into two groups:

- **RX protocols**: Communication between the radio **receiver** (**RX**) and **flight controller** (**FC**)
- **TX protocols**: Communication between the radio **transmitter** (**TX**) and the radio RX

In the following figure, we can see the synchronization of a TX and an RC, as well as their binding with the flight controller:

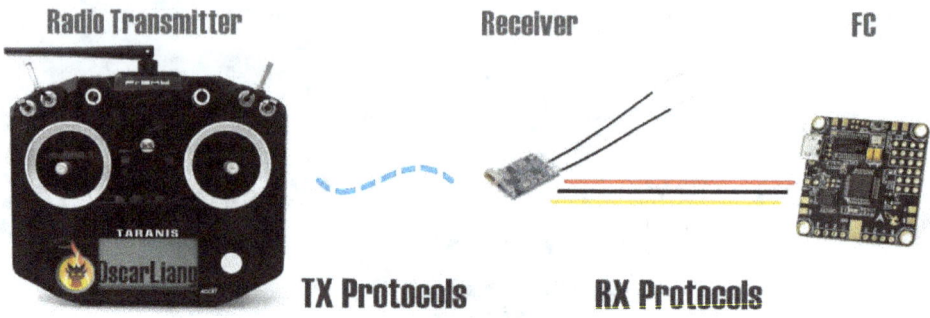

Figure 3.7 – How a TX and RX work with an FC

Transmitter protocols

Wireless communication is a method of communication that's used between a radio TC and an RX. Many manufacturers offer their own TX protocols. Some brands even offer more than one protocol, depending on the hardware used. Here's a list of the most common TX protocols that are used:

- ACCST (Frsky)
- ACCESS (Frsky)
- DSM (Spektrum)
- DSM2 (Spektrum)
- DSMX (Spektrum)
- AFHDS (Flysky)
- AFHDS 2A (Flysky)
- A-FHSS (Hitec)
- FASST (Futaba)
- Hi-Sky (Deviation / Devo)

These protocols have not been explained here because they are not widely used today and are not relevant to this book. These protocols are readily available online if you want a quick overview.

Receiving protocols

The radio receiver communicates with the flight controller via a wire-wired interface. Some RX protocols are standard and found in all radio receivers from the same manufacturer, while others are proprietary to a specific manufacturer. The following are some of the standard RX protocols:

- PPM or CPPM
- PWMSBUS (Futaba, Frsky)
- IBUS (Flysky)
- XBUS (JR)
- MSP (Multiwii)
- CRSF (TBS Crossfire)
- SPEKTRUM1024 (Spektrum DSM2)
- SPEKTRUM2048 (Spektrum DSMX)
- FPort (Frsky)

Pulse position module (PPM)

PPM is also known as CPPM or PPMSUM. The PPM flag is a set of PWM signals that are sent randomly on a single balanced wire.

The advantage of PPM over PWM is that it has one cable for many channels. Therefore, in general, ground lines, power lines, and flag lines should be connected to eight channels.

It is not as accurate or smooth as serial traffic (even small traffic) because channel values do not arrive at the same time:

Figure 3.8 – An image showing the three-pin cable connection of a PPM

Pulse width modulation (PWM)

PWM shares the characteristics of digital and analog signals in that the length of the pulse indicates the servo output or throttle position. Signals are usually between 1000 μs and 2000 μs (microseconds).

This is the most common and original radio control protocol. Since there is no flight controller, the receiver is used to directly control servos and ESCs with standard PWM signals.

The downside to this is that cabling is a problem because there is only one servo cable per channel. So, PPM and SBUS are usually better than PWM when using FC. FC works just as well – if not better – when you can get all the channels on one phone:

Introduction to RCs and the different available options   55

Figure 3.9 – An RC receiver and PWM connection

Serial protocols

A serial protocol is a lossless digital protocol that uses only three wires (signal, power, and ground) for multiple channels. Unlike PPM, which signals in the time domain, the serial protocol is fully digital – that is, it uses 1s and 0s. As the name suggests, the serial protocol requires a serial port on the flight controller.

SBUS protocol

SBUS is the same language that the drone and remote control use to communicate with each other. This is a digital way for the remote control to tell the drone what to do. Communication is easy and efficient because information is sent through a single phone instead of multiple phones. This helps the drone understand and follow the remote control commands so that it can fly smoothly.

Summary

In this chapter, we learned about various command and control interfaces and what they mean for drones. We learned about the diverse types of ground control stations, how they work, and what they mean. We also learned about the protocols and procedures that are involved in RC transmitters and receivers, as well as how to use them. We now have a clear idea of what GCS is and why it is needed.

In the next chapter, you will learn how to design a drone from scratch, how to select each part of the drone, and how to create a successful drone.

4
Knowing UAV Systems, Sub-Systems, and Components

Earlier, we read about the forces that work on a drone, its major hardware, systems, and sub-systems, and the systems that are applicable to the drone. Now, we will go ahead by learning about the development of these hardware systems. In this chapter, we will learn about the major concepts, terminologies, and laws that are helpful in the development of drones as a whole. This would let the user design their own system as per the specifications given.

In this chapter, we will learn how to select the best components for the drone as per their specification requirements and how they can be integrated to develop a drone. We will go through the following topics to help you understand the concept in detail:

- Understanding and conceptualizing the system
- Designing the propulsion system
- Designing and developing the airframe
- Developing the control system
- Selecting the navigation and communication system

Technical requirements

In this chapter, the major technical requirements are basic high school physics, knowledge of the resultant forces that act on the body, and basic electrical engineering and calculations. It would be beneficial to have an understanding of circuit voltage, current, power, batteries, and so on. Experience of handling electrical instruments is a plus.

Understanding and conceptualizing the system

Understanding and conceptualizing helps us to build every kind of drone for its dedicated application and purpose. As we saw in *Chapter 1* the major applications for which a drone can be used, we now see how drones can be conceptualized and developed.

A drone is basically a platform that can fly into the air with a certain amount of payload capacity, which we have already studied with **maximum take-off weight** (**MTOW**). Thus, a drone platform can be helpful in transforming a drone for different applications by applying the permissible payload to it and optimizing its weight, endurance, size, and power.

When we are conceptualizing a system, we need to think about and decide on the following things:

- **The application**: The purpose of this is to know the application and purpose for which the drone is built, for example, an urban mapping by a LiDAR payload, a land survey using an RGB camera, crop spraying/seeding in agriculture, day and night surveillance, or a high endurance tether drone. Knowledge of the application would help you to conceptualize the system in a better way in terms of airframe design and selection and placement of components. You will be better aware of the kind of situation and the work the drone has to go through; thus, the drone can be customized in that way.

- **Endurance (flight time)**: Endurance is the key factor that has to be considered while conceptualizing the drone. Long endurance helps you to fly longer and do more productive missions without landing frequently for battery replacements. However, the demand for the application is still regarded.

- **Range**: The range of the drone is the parameter that says how far it can go with the communication link still connected to the ground control station. This parameter depends on the application we are going to serve. For a photographic drone that does not need to go to large distances, a range of 1 to 1.5 km is enough, but for a survey and surveillance drone required to go long distances, a range of 15-20 km is also mediocre. The big drones that come in the **high altitude long endurance** (**HALE**) or **medium altitude and long endurance** (**MALE**) category have ranges from 200-300 km. Hence, range is also an important factor to decide on.

- **Max payload weight**: This parameter gives you the maximum permissible weight of the payload that the drone can safely take off with. Since the drone has its own weight, now it's the weight of the payload that is variable and should be fixed to the maximum. This would help us to select the payload with the desired weight only to attach to the drone.

- **Features and functionalities**: This point determines what extra features we want to add to the system apart from the traditional flying bird. These features include navigation lights and ADSB in/out. These features are not mandatory features but depend on a drone based on the local laws or customer requirements. Compliance with the country's laws plays a vital role in setting the operational features of the system. Some countries require permissions to fly BVLOS, while some do not. In some countries, such as the USA, remote ID is mandated, while some countries are developing compliance for UAV operations. We can say that the features and specifications should be modified as per local regulations and compliance.

By deciding on the above factors, we can decide step by step the basic components of the drone that we will see in the coming topics.

Designing the propulsion system

The propulsion system or power train is the most important system in the drone. It's the backbone of the drone due to which the drone takes off in the air safely, performs missions, gives the required endurance, and is able to make sharp maneuvers and withstand high winds during the course of action. As we studied in an earlier chapter that the weight of the drone is key to choosing its propulsion system, we can look ahead for the approximate weight distribution of different components for the approximate total weight of the drone.

The propulsion system of the drone consists mainly of motors, propellers, ESCs, power distribution boards, and a battery. These are the key components whose weights play a key role in deciding on the MTOW of the drone. Let's have a look at these components and a methodology to decide on the best components based on the working experience of developing a multicopter.

Let's consider a drone whose total weight is estimated to be 4 kg with payload and whose required endurance is a minimum of 60 minutes. We can divide the approximate weight distribution of the components as follows:

S. No	Components	Weight, as a percentage
1.	Airframe	15%
2.	ESC + motors + propellers	13%
3	Battery	40%
5	Payload	15%
6	Avionics	10%
7	Miscellaneous and future	7%

Table 4.1 – Approximate weight division of the components

The preceding table describes the approximate weight distribution of the components used in the drone. This is purely based on personal experience of developing multicopters and is not a hard and fast rule to follow. From the developmental experiences, the weight ratios tend to follow this trend. Hence, this can be changed based on the user's requirements and application.

Selection criteria for motor and propeller combination

We have already studied in the previous chapters that the motor is used to provide rotation, and based on the rotation, the propeller produces the required thrust to pull the drone into the air. Here, we will see how to select the best power train for motor propellers that would give us the required amount of thrust with less weight, less power consumption, and also a small size.

The tradeoffs in the selection process are as follows:

- **Weight versus power**: A lesser-weight motor to be chosen that consumes less power, providing the required amount of thrust.
- **Size versus power**: Smaller-sized propellers with the motor selected above consume less power but provide an equivalent amount of thrust.
- **Thrust-to-weight ratio**: This is a point to note that in any combination of motors and propellers, the thrust requirement should not be compromised by choosing a smaller size and a low-weight component. Please make sure that the combination produces the required thrust, hence proper focus is required when selecting these combinations.

Thrust-to-weight ratio

The **thrust-to-weight ratio** (**TWR**) is the fundamental ratio in the design and operation of multicopters. The TWR influences the drone's flight dynamics and its capabilities in application. The TWR, as the name says, is the ratio of the total thrust generated by the propulsion system to the total weight of the multicopter.

Importance of TWR

The TWR is important for predicting the multicopter's capability to ascend, maneuver, and carry payloads effectively with responsiveness and stability. This is crucial for developing drones for applications such as aerial photography, surveillance, and recreational flying. A low TWR can make a drone sluggish in performance, making it difficult to execute sharp maneuvers, while an excessive TWR makes a drone compromise its stability.

The ideal TWR for general purpose

For normal multicopters, a typical TWR is typically 2:1 to 3:1. This helps to seek a balance between agility and stability, which allows a multicopter to go through a variety of tasks efficiently. It provides enough thrust for quick ascents and dynamic maneuvers without sacrificing overall control and stability.

Calculating the TWR

The TWR is calculated using the following formula:

TWR – Total thrust produced/Total weight

Here, the total thrust is the combined output of all the propulsion units (motors) in the multicopter, and the weight is the total mass of the aircraft, including any payload.

The differences between a high TWR and a low TWR are shown here:

- **High TWR**: A multicopter with a high TWR, that is, a TWR >3:1, is characterized by a strong thrust relative to its weight. This TWR is appropriate for applications that require rapid acceleration, sharp maneuvers, and fast ascents, such as racing drones.
- **Low TWR**: Alternatively, a multicopter with a low TWR, that is, a TWR <2:1, might be more stable but would lack the agility required for dynamic flight. Multicopters with a lower TWR are usually deployed in applications where stability is a prime requirement, such as aerial photography or surveillance.

The following are the requirements for achieving an optimal TWR:

- **Motor and propeller selection**: Choose motors and propellers that fit the planned application. High-quality, powerful motors with compatible propellers result in efficient thrust production.
- **Battery choice**: Select batteries with the required capacity and discharge rate to supply the required power for optimal thrust.
- **Weight management**: Optimize the overall weight of the multicopter by using lightweight materials for the frame and components, creating an equilibrium between structural strength and weight reduction.
- **Efficient electronics**: Utilize efficient electronic components, including **electronic speed controllers** (**ESCs**) and flight controllers, to enhance power management and overall performance.

Characteristics of motors to be used

Since we are developing a drone, we cannot use any motors that are available on the market. The recommended motors in a drone are **Brushless DC (BLDC)** motors, which are highly efficient with a lower weight and a smaller size and consume less current. The motor should also produce enough torque to rotate the propellers to produce the desired thrust. Here are the few specifications that we should know about the motor as a drone engineer:

- **KV rating**: KV ratings of BLDC motors specify the RPM at which the motor rotates with every volt applied to it under no load conditions. Basically, in simple terms, the KV rating states the RPM per volt for a BLDC motor. If a motor is 100 KV, it means that the BLDC motor will rotate 100 times if 1 Volt is applied to it. In real-time conditions, there would be some load on the motors due to the propellers and air resistance in the rotation, hence the motor would not be able to reach the idle RPM but would come close. Due to the load, the concept of torque comes into the picture, which we will see later.

Hence, while deciding on the motor, the battery configuration should also be kept in mind, specifically which battery these motors are to be run on, that is, 6 cells, 8 cells, or 12 cells. For a 100 KV motor and a 25 Volt battery, the RPM at max voltage would be 25 x 100 = 2500 RPM.

- **Motor torque**: Torque is basically the force at which the motor rotates the propeller against wind resistance. In other words, it's the force of the motor needed to overcome the load applied. Hence, bigger propellers and heavy-load applications use motors with higher torque, and smaller-load applications use motors with lower torque and higher RPM.

  The torque and KV of the motors are inverse to each other. The higher the KV, the lower the torque, and vice versa. Motors with a higher KV would have a higher RPM at a particular voltage and less torque, hence they take a smaller propeller at a higher RPM to produce the thrust.

  Motors with a lower KV have a lower RPM and high torque, hence they run slower but use bigger propellers, so the amount of thrust produced will be the same.

- **Operating voltage**: This is the most important parameter in motor selection. We need to be sure that the battery we are using is 6s, 8s, 12s (and so on) and select the motor that operates at this voltage. Choosing a lower operating voltage motor with a higher battery voltage will burn the motor, hence a motor should be selected based on the proper voltage of the battery we are choosing.

Motor specifications and thrust charts

Motor specifications include the motor's mechanical drawings as well as its specifications, whereas thrust data charts are particularly important and give you performance details about the motor with different compatible propellers. These charts are given by the motor manufacturers and contain precise details about the thrust, power torque, and other parameters that are helpful in observing the motor performance and selecting the right motor. The same data can be deduced using the thrust bench locally for any motor.

To select the motors for our application, we can go through different manufacturers' motor specifications and thrust bench data and perform a comparative analysis of these motor thrust data. We can go through the following parameters:

- **Voltage**: Since we are building a system with a 6s configuration, we would go ahead with motors that support a minimum of 6s (22.2 nominal voltage). Since the following motor works on a 6s configuration, we can go ahead with it.

- **Thrust**: We would observe the thrust produced by these motors, which should be nearly equal to the weight of the drone (keeping a TWR of 2:1), and select the propeller combination that produces that thrust.

- **Weight and size**: Considering the above factors, we would look for the combination that produces the required thrust with the least motor weight and the minimum propeller size. This would help us to reduce the size and weight of the drone.

- **Current consumption**: Since we need high endurance, we need to look for the combination that provides us with the same thrust, less weight, a smaller size, and less current consumption.

The previous points are the four main points to be observed while selecting the power train for your drone. The thrust bench data might not be accurate due to the real-time scenario that includes heat losses and other environmental factors.

By considering the previous factors, we will study the thrust chart for a T-Motor Antigravity 5006 300 KV Motor, which has various size propellers at different throttle percentages and reflects its performance.

A product drawing of the motor helps the designer to extract the following mechanical information:

- Mounting hole diameter
- Distance between holes
- Diameter of motor and thickness

This information helps the designer to provide the mounting structures in the airframe design and better optimize it.

The following are the specifications and mechanical drawings of the selected motor:

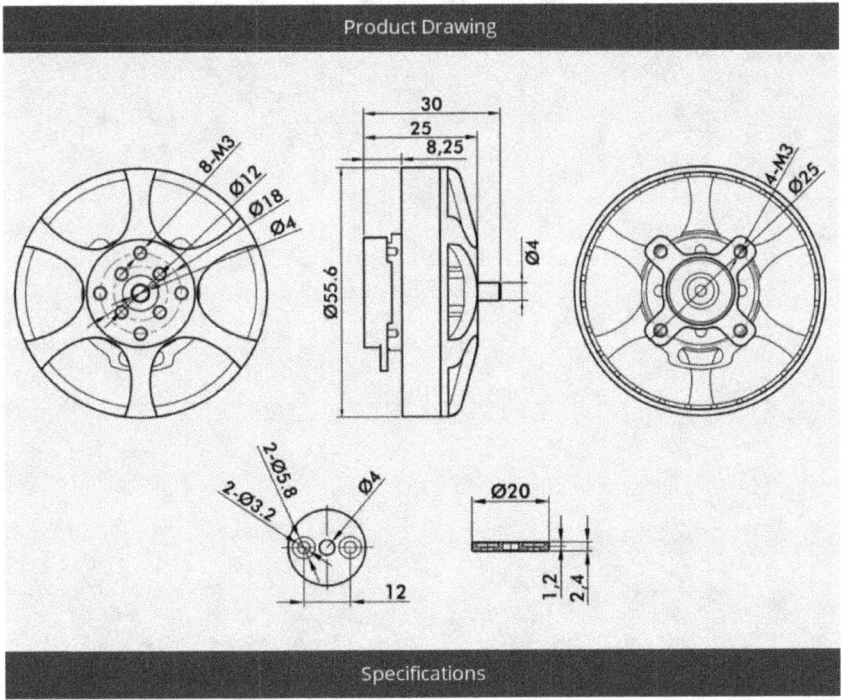

Figure 4.1 – Drawing of MN5006 300 KV Motor (source: T-Motor)

The following is an image of the motor specification provided by the manufacturer T-Motor that clearly states the physical properties and the necessary electronics recommended for use with this motor. This helps avionics engineers calculate the weight of the drone and overall power consumption by the propulsion system:

KV	300	Rated Voltage (Lipo)	4-6S
Idle Current (22V)	0.6A	ESC Recommendation	AIR 40A
Peak Current (180s)	20A	Propeller Recommendation	P17-18"
Max. Power (180s)	500W	Motor Weight (Incl. Cable)	108g
Internal Resistance	125mΩ	Package Weight	170g

Figure 4.2 – Specifications of MN5006 300 KV Motor (source: T-Motor)

The following is an image of the manufacturer's thrust bench data, which will help us select the propellers for our 4 kg drone that we decided on earlier. The datasheet shows us the motor, along with the propellers at different throttle percentages, which gives us the current consumed along with power, RPM, and torque. It also gives the maximum temperature to which the motor is heated up in the process:

Type	Propeller	Throttle	Voltage (V)	Current (A)	Power (W)	RPM	Torque (N*m)	Thrust (g)	Efficiency (g/W)	Operating Temperature (°C)
MN5006 KV300	T-MOTOR P17*5.8" CF	40%	23.70	1.63	39	2481	0.10	501	13.02	67
		45%	23.67	2.21	52	2790	0.13	640	12.24	
		50%	23.64	2.88	68	3103	0.15	784	11.5	
		55%	23.61	3.71	88	3410	0.19	942	10.75	
		60%	23.57	4.66	110	3689	0.22	1119	10.18	
		65%	23.54	5.68	134	3947	0.25	1289	9.64	
		70%	23.50	6.78	159	4203	0.29	1454	9.13	
		75%	23.46	8.00	188	4442	0.32	1631	8.69	
		80%	23.42	9.28	217	4664	0.35	1798	8.28	
		90%	23.32	12.31	287	5105	0.43	2186	7.62	
		100%	23.22	15.41	358	5491	0.50	2538	7.09	
	T-MOTOR P18*6.1" CF	40%	23.68	2.04	48	2410	0.14	629	13	82
		45%	23.64	2.80	66	2702	0.17	792	11.96	
		50%	23.61	3.67	87	2988	0.21	968	11.18	
		55%	23.57	4.75	112	3264	0.24	1162	10.38	
		60%	23.53	5.91	139	3525	0.29	1366	9.82	
		65%	23.48	7.25	170	3762	0.33	1572	9.24	
		70%	23.43	8.69	204	3996	0.38	1777	8.73	
		75%	23.38	10.22	239	4219	0.42	1980	8.29	
		80%	23.33	11.89	277	4424	0.46	2183	7.87	
		90%	23.21	15.70	364	4807	0.55	2618	7.19	
		100%	23.09	19.63	453	5131	0.63	2996	6.61	

Figure 4.3 – Test report of MN5006 300 KV Motor (source: T-Motor)

Motor and propeller selection

In the above chart, we can see that for an 18*6.1" CF propeller at 50% throttle, thrust equals 968 grams per motor. Since we are developing a quadcopter, the thrust produced by all four rotors is 968 x 4 = 3872 grams, which is nearly equal to 4 kg (weight of the drone), keeping the TWR at 2:1 for general-purpose applications.

The 17*5.8" CF propeller is less efficient then 18*6.1 CF propellers. So we can go ahead with the selection of 18*6.1 CF propellers.

The current consumed per motor at 50% throttle is 3.67 amperes, while the current consumed by all four motors at 50% throttle: 3.67 x 4 = 14.68 amperes. Hence, the total thrust produced by the motors is sufficient for hovering at 14.68 amperes, which we got from the specification chart. This data is sufficient for us to go ahead with this combination, as the operating temperature does not get extremely hot in this scenario.

Hence, we can say that this motor suits our requirements and proves to be efficient as per the datasheet. However, this is not the ideal case; in a practical scenario, there would be many other factors such as wind, speed, flying method, etc., that would play their role, and the data presented above might be slightly affected.

The next step is to calculate the flight time with the combination to achieve a good flight time. Let us consider a proper battery for this application.

Battery selection

The battery is the powerhouse of the drone. As we discussed in an earlier chapter the different types of the battery, there are certain ratings that we need to keep in mind before selecting the right battery for our application.

We have now got the current consumption of the drone from the above charts. The total current consumption will always be 5% higher due to other avionics and environmental effects such as wind, so let's keep the current consumption at 14.68 + 5% = 15.414 amperes, which means that the total current consumption of the drone can be considered as 15.414 amperes. We will use this as a base for all calculations.

The following are things to keep in mind when selecting a battery:

- **Weight**: As we have a flying system, it is always advisable to keep the weight as low as possible. We should always look for the battery that has a high energy density and a low weight, which would give us better efficiency and higher endurance.

- **C ratings**: As per battery university, the charge and discharge rates of a battery are governed by C ratings. C ratings determine the safe discharge of the battery, i.e., the number of amperes that a battery can safely give the system without getting damaged. It is mentioned as 5C, 15C, 22C. It is calculated as follows: let us say that the capacity of the battery is 16Ah, that is, 16000mAh, the C rating is 5, and safe discharge is equal to Ah x C, so 16 x 5 = 80 amps. Thus, the battery can give 80 amps continuously without getting damaged, though our application should be consuming less than that.

We can choose any battery chemistry from among lithium polymer, lithium ion, and solid-state lithium polymer if they satisfy the above ratings considering the weight and energy density of the battery.

Calculating the flight time (endurance)

The flight time is calculated using a simple fundamental equation: energy stored divided by energy consumed.

For our application, the total current consumption came out to be 15.5 Amps (rounded off). If we select a battery of 16 AH, it would satisfy our endurance requirement with minimal weight.

Formula to calculate flight time

The flight time (T) of a multicopter can be estimated using the following simplified formula:

$$\text{Flight Time} = \frac{\text{Capacity(Ah)} \times 60}{\text{Ampere Draw}}$$

where flight time is to be calculated, capacity in Ah = 16Ah or 16000mAh, and ampere draw is 15.5.

Putting the values into the formula gives 61.93 minutes.

Hence, we have calculated the flight time of the 4 kg drone to be 61.93 minutes.

This is how we can select the right propulsion train and calculate the flight time of the system. Note that the flight time is an approximate, ideal flight time. In reality, flight time is impacted by temperature, humidity, flying altitude, wind, and many more factors due to which endurance can vary.

Designing and developing the airframe

An airframe is the skeleton of the drone that holds the avionics and other components together tightly and stiffly. It's the structure that actually moves into the air and gets the effect of the environment; hence, an airframe should be rigid, strong, and lightweight. We can either design our own airframe using the different software available and following the composite manufacturing process or we can use the airframes available on the market.

Whether we are designing the airframe or getting the airframe, we need to consider the following points that would be helpful in giving us an idea of how much load the drone can carry and how maneuverable it can be:

- **Professionally engineered**: When an airframe is developed in a laboratory, certain materials are selected, such as carbon fiber, glass fiber, and so on, and have to go through the manufacturing process of cutting drilling. The material, that is, carbon fiber, is laid up in molds using resin with certain layer sequences and kept for curing. This makes the fiber hard and makes it take the shape of the mold. Hence, the desired shape comes out and is then engineered using machining.

  This is the process of building an airframe if you have designed it in-house and have some unique concepts regarding it. However, for a beginner build, there are quadcopter/hexacopter airframes available on the market to start the build.

An airframe should be smoothly finished internally and externally without any rough surface and equivalently sized across all axes.

- **Configuration and size**: The general term of measurement of an airframe is known as the wheelbase. The wheelbase is known as the diagonal center-to-center distance between two motor mounts. The propellers should perfectly fit into the airframe without colliding with any parts of the airframe or themselves in the air. There shouldn't be any loose ends in the airframe. The airframe should also be balanced under all configurations with a proper center of gravity and center of mass.

- **Mounts and fixtures**: Mounts and fixtures are the hardware on which avionics such as motors, ESC, and other parts are installed on the drone in the desired manner. These mounts and fixtures for the motors are designed and developed with holes and fixtures for the placement of PDBs, ESCs, and flight controllers. Payload mounts and spaces have to be maintained for free movement of the payload. The details for mounting hardware, such as screws, glues, nuts, and bolts, should be proper.

- **Material**: The material for manufacturing the airframe should be lightweight and sturdy with good strength across all axes. Composite materials such as carbon fiber and glass fiber or Kevlar are widely used for manufacturing airframes due to their lightweight and excellent strength properties. These materials should also be passive towards the environmental conditions and not be subject to wear and tear in the long run.

- **Strength**: The airframe should have adequate strength to handle the load of force produced by the motors and gravity. Vulnerable areas such as arm joints and motor mount joints must be strong and able to take the load mid-air under sharp maneuvers.

Developing the control system

In this section, we will cover control system development.

Electronic speed controllers and flight controllers

An **electronics speed controller** (**ESC**) is a small device that comes between the **flight controller** (**FC**) and the BLDC motors. An ESC, as the name suggests, controls the speed at which the BLDC motor rotates as per the signals received from the flight controller. Each motor requires one ESC to control its speed and is connected to the flight controller. The FC simultaneously has **pulse with modulation** (**PWM**) pins to which the ESC is connected. The FC gives the PWM/DShot signal to the ESC as per the commands given by the GCS RC to run the motors. The ESC gets power from the battery and signals from the flight controller. The following diagram illustrates this setup:

Developing the control system 69

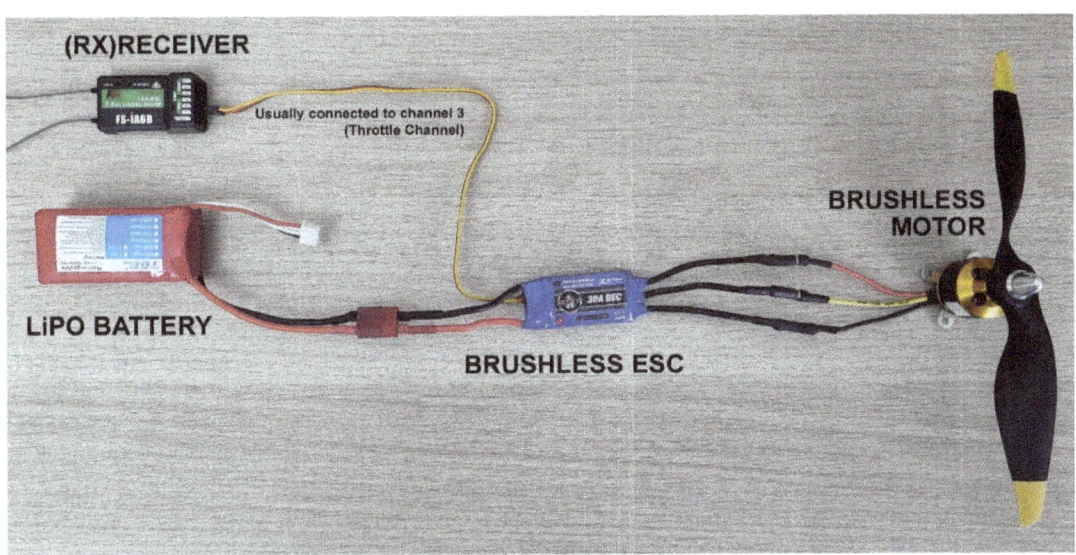

Figure 4.4 – The workings of an ESC

The ESC is connected to the battery and supplies the battery voltage to the motors. The ESC is a kind of oscillator or pulse generator that passes a variable duty cycle and is commonly used as a speed controller for motors. This is controlled by the FC via the PWM signals. These PWM signals generated by the FC to operate a BLDC motor are fed into the ESC. The value of a PWM signal varies between 1000 to 2000 microseconds, where 1000 represents the lowest voltage level and 2000 is the highest voltage level that has to be supplied to the motors. This voltage value depends on the value of the PWM being supplied to the ESC. The flight controller has special motor driving algorithms due to which roll pitch and yaw are controlled based on the speed of the motors.

The factors to be considered while choosing the ESC for the drone are as follows:

- **Current rating**: The current rating is the first and most important thing that has to be checked. The current rating of the ESC is how much current can safely pass through it without getting burnt. The current is pulled by the motor to work, which passes through the ESCs. For this, the correct current rating of the motor-prop combo must be known. In our earlier example from the motor sheet, we can see that each motor can draw a current of 19.6 amperes, hence the ESC should be a minimum of 10% above this rating.

  There are two types of current ratings for the ESC:

  - **Continuous current**: Continuous current is the maximum stable current that ESC can pass through it with spikes and fluctuations.
  - **Burst current**: Burst current is the maximum amount of current for a short period of time, that is, for 10 seconds, without damaging the ESC. The burst rating of the ESC is helpful in managing conditions such as heavy wind and sharp maneuvers where motors have to spin faster and demand extra current.

- **Operating voltage ratings**: Operating input voltage refers to the maximum input voltage of the battery that the ESC can handle and on which the motor works. Many ESCs support 4S to 6S (S denotes the number of cells) batteries while others support 12S to 14S batteries. It depends on the application as to what ESC we require. For the 4 kg configuration that we decided on earlier, a 40-ampere ESC is enough.
- **Weight and size**: The size and weight of the ESC are crucial factors as they impact the weight of the drone and its placement on the airframe. The greater the current rating, the greater the size of the ESC. Some higher current ESCs also come with a heat sink, which makes the size and weight greater.
- **ESC protocols**: Protocols are the communication tools between the flight controller and the ESC. They define how fast communication and time synchronization happen between the flight controller and the ESC. A better ESC is something that understands all major protocols. The general protocols in use include **pulse width modulation** (**PWM**), which is an analog protocol for signal transmission to ESCs. However, newer and more efficient digital protocols have evolved – Dhot and proshot – which are more accurate and faster.

Hence, the ESC is an important part of the drone and is to be chosen based on the current consumption by the motors and the latest protocols.

Flight controller

A flight controller is the most important and crucial part of the drone; it helps to control the drone's movement and stability. Numerous factors must be kept in mind when choosing the right flight controller based on the features and protocols and also based on the IMU and sensors.

A flight controller is the brain of the drone. It primarily controls the functioning of the ESC and motors based on the inputs received from the sensors. A flight controller has onboard sensors, processors, communication protocols, and interfaces for outside communications. The purpose of the flight controller is to stabilize the drone during the flight using its on-board sensors, which consist of the IMU we have discussed previously.

How to choose a flight controller

Since there are various flight controllers available on the market, how can we choose the best flight controller that is suitable for our applications? Based on various factors, the best flight controller can be chosen:

- **Microcontroller**: Every flight controller has a microcontroller that governs all its decisions and the speed of the operations. In the scenario of critical decision-making, we need a flight controller that can make quick decisions and process data quickly. There are many flight controllers that come with different microcontroller boards such as Atmel Mega 644PA 8 Bit, ATMega 328P, STM 32 F4 32 bit, etc. This is similar to how we look at a processor when choosing a laptop, that is, i5/i7/Ryzen.

Hence, the flight controller with the best microcontroller board must be selected if fast processing of the data is required.

- **Firmware**: Firmware is the soul of the flight controller and therefore the drone, so the flight controller must be chosen so that supports and is compatible with trustable firmware. Some key open source firmware widely used today include ArduPilot and PX4. Apart from them, there are beta flight and INAV, which are leading the industry. The firmware code should also be readily available, and modification based on the user should be possible when required.

  Let's see the reputed industry flight:

  - **ArduPilot firmware**: ArduPilot is a reputed open source flight stack that supports multiple vehicle configurations, including multicopters. It contains a brief set of features helpful for both hobbyist and professional use:

    - **Versatility**: ArduPilot flight stack is versatile as it supports various vehicle types, such as multicopters, planes, rovers, and boats
    - **Feature-rich suite**: ArduPilot contains many features, including waypoint navigation, return-to-home, altitude hold, and support for various sensors and actuators
    - **Large and active community**: The user community of ArduPilot is large and active, ensuring progressive development and support

  - **PX4 firmware**: PX4 is a similar reputed open source flight stack designed for unmanned vehicles, including multicopters. It is enriched with its modularity and real-time capabilities. Some of its key features are as follows:

    - **Modularity and adaptability**: The basic design of PX4 is modular, empowering users to modify and adapt it to different vehicles and configurations.
    - **Real-time operating system (RTOS)**: PX4 consists of an RTOS, providing precise and low-latency control. This is particularly useful for applications that demand responsive performance from multicopters.
    - **Hardware integration**: PX4 supports a wide range of hardware components compatible with its stack, providing users with flexibility in selecting compatible sensors, controllers, and other components.

- **Sensors**: Sensors are the primary source collecting the physical information of the drone; they act as the sensory organ of the drone. Nowadays, flight controllers come with built-in sensors such as accelerometer, gyroscope, temperature sensor, and magnetometer, which provide the data to the microcontroller to make the required decision as per the firmware written. Sensors are majorly used for knowing the orientation of the drone and stabilizing it during roll, pitch, and yaw.

The following is a list of the sensors used in different kinds of boards:

- **KK 2.1.5 Flight Controller Board**: MPU6050 gyroscope and accelerometer
- **ArduPilot APM 2.8**: 3-axis gyroscope accelerometer, barometer
- **OpenPilot CC3D**: 3-axis gyroscope accelerometer
- **ArduPilot APM 2,6**: 3-axis gyroscope accelerometer, barometer

- **Communication protocol**: To work with any peripheral devices, such as GPS, range finders, or sprayers, we need a communication protocol that is workable with the peripheral device. To make this happen, we require a flight controller that can support multiple communication protocols such as UART, I2C, CAN, PWM, and so on. We discussed these protocols in our earlier chapters. These protocols are used to transfer the sensor's data back and forth to the flight controller, which makes the required decision with speed and accuracy. A few examples of flight controllers with communication protocols are as follows:

  - **KK flight controller**: Built-in LCD display and four buttons for configuration
  - **ArduPilot APM flight controller board**: PWM, I2C, UART, PWM, PMU, and SBUS
  - **OpenPilot CC3D flight controller**: UAR, I2C, and SBUS
  - **ArduPilot APM 2.6 flight controller board**: UART, I2C, SBUS, CAN, etc.

The abovementioned flight controllers are crucial parameters for choosing the best flight controller based on the application, required communication protocols, and firmware. A few more minor points include the size and weight of the controller and power consumption, which should be extremely low.

Some famous flight controller boards

Earlier, we studied the different features required for the flight controller to be used as per the application. Here is a list of some industry-famous flight controller boards that run in different drones:

- **KK flight controller board**: The KK flight controller is a beginner flight controller with an inbuilt LCS and four buttons to configure the board for flight. This flight controller is built for multirotor, tricopter, quadcopter, or hexacopter and keeps them stable during the flight. The flight controller is based on Atmega644 PA IC coupled with InvenSense MPU-6050 IMU sensors.
- **Pixhawk 2.8 flight controller**: As the unmanned industry has grown, we have started using the same autopilot boards or hardware for multiple unmanned applications. The Pixhawk flight controller is an open source flight controller used to build fixed-wing drones, multicopters, and rovers, both with a change of firmware by using the same hardware. This gave us the chance to build more unmanned vehicles using the same hardware and sensors. Pixhawk can perform pre-programmed autonomous missions using GCS software and multiway point missions using the on-board compass, which makes it an advanced flight controller. The integrated 3-axis IMUs

and on-board logging with ATMega chip make this flight controller more advanced and easier to use for automated missions with dedicated failsafe modes.

- **DJI Naza**: DJI Naza is an advanced flight controller with fully enhanced features that can support up to nine types of multi-rotors. It comes with configured GPS, assistant software for smartphones, and a free ground control station. The flight controller comes with power output fail protection, intelligent operation control, flight modes, and a multiple peripheral communication protocol that helps it connect to output sensors for data and decision-making, in turn making it more unique. DJI Naza is a great board for advanced users who are looking for stable and long-haul flights.

- **Pixhawk Cube Orange**: Pixhawk Cube Orange is the most widely used open source flight controller in the unmanned industry. One of the biggest advantages of using this Pixhawk flight controller is its open source design that gives even the minute details of the hardware, which in turn allows access to the detailed information of the hardware and the design. This is coupled with open source drone firmware such as ArduPilot or PX4, which are the two most advanced and widely used firmware with the Pixhawk Cube Orange for running any unmanned system. The Pixhawk Cube supports multirotor, fixed-wing, and VTOL drones, rovers, and submarines. With a change of firmware, this can be equipped in any vehicle. What makes Cube Orange advanced is that it comes with the STM32H7 microcontroller and one failsafe co-processor. It has high connectivity options for external hardware and peripheral interfaces such as SPI, I2C, PWM, and CAN. It has a redundant power supply and a microSD card for data logging.

As a result, we can say that the listed flight controllers are modern autopilot systems that can be used in building a drone for a specific application. We also understood how to select peripheral devices and autopilot systems for our drone in the preceding topic.

Selecting the navigation and communication systems

Navigation and communication systems are the most important systems in the drone. The communication system helps to command and control the drone over longer distances, and the navigation system helps the drone decide its north and guides the system to the waypoints in an autonomous mission.

Communication systems

A communication system basically consists of a radio modem and an antenna that is connected to the ground control station and the drone, which are connected to each other. This system helps to program the drone for autonomous missions and command the drone mid-flight to perform specific tasks. Let's see the parameters at which a communication system exists (examples of a radio modem is RFD900 433Mhz.):

- **Range of drone**: Range is a crucial factor that helps the engineer to decide and integrate the communication system. As the range requirement increases, the radio modem and antenna requirements also increase.

- **Functionality**: Functionality is another important factor that plays an important role in the selection of the communication system. We need to see whether the application requires a video transmission system along with the command and control, and then such radio modems have to be chosen.

- **Working protocols**: To make the communication compatible with the system, it should match the working protocols of the drone flight controller and should have sufficient ports for it. Some of the working protocols of the communication systems are as follows:

  - **Serial UART**: Serial UART is a general traditional UART protocol used to make a serial connection to the drone. It directly connects with the flight controller serial connection.

  - **Network-based communication system (smart radios)**: There is a certain communication radio device with a smart radio and an IP-based network that is an advanced communication device. In these devices, the network is formed between two or more devices, and data is shared among them. These devices are new-age devices that work on internet protocol-based network communication, which is helpful in sharing audio, video, and command and control. These devices work under many modes, such as master-slave mode and mesh mode. These are different network topologies that are helpful in governing the type of network among them. Some of the examples of such devices are Doodle reduce microhard and silvus radios, which are extensively used in the drone world.

- **Power consumption**: Before choosing a radio modem, it must be kept in mind that the radio modem should consume less power from the drone since it may directly impact the flight time of the system, hence the radio modem should consume as little as possible, even at the farthest distance from the ground control station. Therefore, it's better to ensure and check the data sheets before purchasing the radio modem.

- **Size and weight**: As we all know drones are used in remote areas, are operated by a limited number of people, and also have a weight constraint, as we have seen in the earlier topics, before choosing a radio, we should make a note that it should be as lightweight and small in size as possible so that it does not add extra weight to the drone. The system must also be less complex so that it becomes easy for transportation and assembly on the GCS side and consumes less time.

- **Legality on frequency and power**: It has been seen that in different countries, there are different regulations on the maximum power that can be emitted. Certain bands of frequency are also permitted for use without any license. To use any other spectrum band apart from the regulated one, a license has to be obtained from the government of the state. For example, 2.4 GHz and 5.8 GHz are unlicensed frequencies used for Wi-Fi communications that do not require permission from the government in most countries, such as India. Before selecting any radio, the output power and frequency must have been checked as per local regulations.

Integrated GCS system with communicating devices

Integrated GCS systems are handheld devices that contain a joystick, a screen as well as a radio modem integrated into one housing and powered by a small battery. These are small handheld systems that are used as drone controllers and have the functionality to operate through software and manually by the joystick. They also have several buttons to configure different functionalities or to configure the controller. They have an integrated radio modem that helps the drone achieve long ranges, along with command and control and live video transmission functionalities, which makes them lightweight and easily transportable. The following are some examples of such devices:

- **Herelink**: Herelink is manufactured by CubePilot for the unmanned ecosystem. It is a small compact handheld device with a 5.5-inch screen, an integrated Android version of QGroundControl GCS, and an inbuilt radio modem that works on a 2.4 GHz bandwidth with joysticks and buttons. It supports RC data telemetry data and live video transmission up to the manufacturer's claimed range of 15 km and also allows video streaming over networks. This is the best option for new drone pilots who require a small-scale lightweight integrated GCS for live video streaming and long-range application. This works as a complete GCS system for any type of manual and autonomous mission. More details are available on the Herelink website, which guides you to configure the system on the drone.

- **Siyi AK28**: SIYI AK28 is a handheld GCS similar to Herelink, which we saw earlier. This transmitter is manufactured by SIYI specifically for agriculture drones. The transmitter is equipped with a 5 km range radio modem with a 7-inch display running its own GCS software. This runs for 10 hours on a single charge. It is an easy and more efficient integrated GCS that comes with RC data, telemetry data, and live video transmission. This device is also easy to configure with 4G networks and provides effective mission planning and control for agriculture spray drones.

Some examples of famous communication systems are as follows:

- **RFD 900 radio modem**: RFD 900 is a serial communication protocol-based radio modem that is helpful in long-range communication between the GCS and the drone. The radio is an ISM band radio that operates in the frequency range of 900 to 928 MHz. Since this is telemetry based on serial protocol, it provides the command-and-control interface directly to the drone via GCS without any control of the remote. It does not have video streaming capability due to the lower data rate and serial protocol. It is a small lightweight radio that is license free in Australia, Canada, the USA, and New Zealand.

- **Doodle Labs radio**: Doodle Labs radio is a family of mesh radios designed for industrial applications. They are smart radios known for their long range, high throughput, and low **SWaP (size, weight, and power)** requirements. Doodle Labs radios are not general serial communication radios such as RFD 900. They are IP-based radios that form a network and transport MAVLink data on radio frequency signals under this network.

Doodle Labs radios are used in a variety of applications:

- **Drones and aerial vehicles**: Doodle Labs radios are used to provide long-range communication for drones and other aerial vehicles. This allows them to fly over long distances and stay connected to the ground station.
- **Microhard and UAV radios**: These are similar companies to Doodle Labs that develop smart radios or drones and other communication applications. Both radios have their own operating systems and software that are different from each other in terms of performance and power consumption. Both fall under the category of smart radios that are capable of transmitting control and communication signals and video to long ranges.

Selection of GPS and navigation system

As we have seen in *Chapter 1* about the use of GPS and a compass in the drone, here we will see how we can select the GPS and compass for our drone. As the drone industry is a sunrising industry, many manufacturers providing GPS and a compass together are available. A **global navigation satellite systems** (**GNSS**) device can be relied on for better use in automotive and defense sectors, but its practical data that can be used is divided into the following categories:

- **Location**: This determines your position in the world
- **Navigation**: This identifies the best route from one location to another
- **Tracking**: This monitors an object's movement in the world
- **Mapping**: This creates a map for a particular area
- **Timing**: This computes precision timing

Since these benefits of the GNSS device can be incorporated into the drone for various applications, the use of GPS makes it efficient. The following is a list of criteria that must be met before selecting the right navigation module.

Number of GNSS satellite constellations

We all know that there are different GNSS satellite constellations sent by different countries that are used to provide GNSS services. For example, we have the following:

- GPS from the USA, whose frequency bands are L1 (1575.42 MHz), L2 (1227.60 MHz), and L5 (1176.45 MHz)
- Glonass from Russia, whose frequency bands are L1 (1598.0625-1605.375 MHz), L2 (1242.9375-1248.625 MHz), and L3 (1202.025 MHz)
- Galileo from the European Union, whose frequency bands are E1 (1575.42 MHz), E5 (1191.795 MHz), E5a (1176.45 MHz), E5b (1207.14 MHz), and E6 (1278.75 MHz)

- Beidou from China whose frequency bands are B1I (1561.098 MHz), B1C (1575.42 MHz), B2a (1175.42 MHz), B2I and B2b (1207.14 MHz), and B3I (1268.52 MHz)
- QZSS Japan, whose frequency bands are L1 (1575.42 MHz), L2 (1227.60 MHz), L5 (1176.45 MHz), as well as L6 (1278.75 MHz)
- IRNSS/NavIc from India, whose frequency bands are L5 (1176.45 MHz) as well as along the S-Band (2492.028 MHz)
- GPS, Glonass, Galileo, and Beidou form the global navigation systems, whereas QZSS and IRNSS/Navic form the local navigation system for a particular country.

The following table shows different GNSS constellations with their names, operators, and other useful details:

Comparing GNSS constellations

	Operator	Coverage	Altitude (km)	Satellites in Orbit
GPS	US Space Force	Global	20,180	31
GLONASS	Roscosmos	Global	19,130	24
Galileo	GSA and ESA	Global	23,222	26
BeiDou	CNSA	Global	21,528 (MEO satellites) 35,786 (GEO and IGSO satellites)	48
QZSS	JAXA	Regional	32,000 (perigee) 40,000 (apogee)	4
IRNSS/NavIC	ISRO	Regional	36,000	8

Figure 4.5 – GNSS constellation comparison table

A good GPS module supports the majority of these satellite constellations and helps to attain good accuracy. For example, most of the GPSs use all four global navigation systems, such as Ublox and Here3.

Communication protocol

Like other peripheral devices, GPS devices require a communication protocol to communicate with the flight controller, so we need to look at what communication protocols the GPS works with and whether they are compatible with the flight controller.

Weight, size, and power consumption

A peripheral unit must be lightweight and have a small size and low power consumption with better accuracy.

Some examples of GNSS GPS devices manufactured around the world and extensively used in drones are as follows:

- **Here3**: Here3 is the most famous and widely used GNSS device for open stack ArduPilot and PX4. Here3 serves to be reliable and gives you the position of the drone in latitude and longitude with an accuracy within centimeters. This proves to be more accurate in getting the exact location of the drone. The device supports the **CAN (controlled area network)** protocol with high data rates and upgradability, due to which the signal quality is better and less immune to external noise. The device runs off a microcontroller board with a built-in **IMU (inertial measurement unit)** for advanced navigation needs.

- **Drotek DP0601**: Drotek DP0601 is a **real-time kinematics (RTK)**-based GNSS device that supports GPS, GLONASS, Beidou, Galileo, and QZSS signal reception, gives centimeter-level precision, and comes with built-in support for standard RTCM corrections. It has built-in communication protocols that support both UART serial protocols and SPI, making it suitable for use with multiple devices.

- **Synerex MDU-2000**: MDU-2000 is an RTK-based GNSS receiver that comes with a dual antenna for centimeter-level accuracy. The device does not get impacted by external electromagnetic fields due to its dual antenna technology that provides its true heading even in high magnetic field interference.

Hence, we got an idea of how we can select a GNSS navigation device for our drone with the above mentioned criteria and saw a few examples of the GNSS devices that are widely used in the drone industry to aid the GNSS for the drone.

Summary

In this chapter, we have learned about the theory and selection of the key parts of the drone and the criteria that can be selected for them from the propulsion system to the GNSS device. This is the key concept of making a blueprint for the drone on all the parts available.

In the next chapter, we will focus more on the assembly and integration of these parts and the configuration software used for flying.

Part 2: System Conceptualization and Avionics Development

Now that we have been introduced to the anatomy of a drone, that is, a quadcopter, in this part, we will study drones in a more detailed manner, including how each part works and how the quadcopter works. We will learn how variations in each part influence the working and dynamics of the quadcopter in mid-air. This part will be more detailed and analytical, with mathematical concepts and terminology to develop the system as a whole in software and hardware.

This part has the following chapters:

- *Chapter 5, Sensors and IMUs with Their Application in Drones*
- *Chapter 6, Introduction to Drone Firmware and Flight Stacks*
- *Chapter 7, Introduction to Ground Control Station Software*
- *Chapter 8, Understanding Flight Modes and Mission Planning*

5

The Application of Sensors and IMUs in Drones

In this chapter, we will learn and understand the major sensors that are used in drones to lift into the air and help their systems fly and navigate to waypoints. With the help of these sensors, the drone can fly in various modes and various patterns in manual and autonomous modes. We will learn about the working of these sensors in this chapter, how they interface, and what output they give. We will also learn about the impact they have on the drone's system. At the end of this chapter, we will be diving deep into the anatomy of a drone, which helps us to understand how it works.

We will cover the major sensors required in the drone system under the following topics:

- The inertial measurement unit and its role
- The barometer and its role
- GPS and magnetometer and their roles
- Voltage and current sensors and their roles
- Sensor fusion and state estimation

By going through the above topics, we will briefly be able to understand what the role of each sensor is, and this knowledge will be helpful in deriving the necessary equation and programming of the drone in that aspect.

The inertial measurement unit and its role

The **inertial measurement unit** (IMU) is the most important part of the drone as it helps the drone calculate its linear and angular acceleration; it also helps to measure the attitude angles across roll, pitch, and yaw maneuver, helping the autopilot make decisions and guide its motors and actuators to produce the desired output for level flight.

Composition of an IMU

The IMU is a critical component in multicopter flight control systems. It integrates multiple sensors to provide comprehensive information about the vehicle's motion and orientation. The IMU typically includes accelerometers, gyroscopes, and sometimes magnetometers. The integration of these sensors is essential for accurate flight control, stability, and navigation.

Components of the IMU:

- **Accelerometers**: These measure linear acceleration along different axes (X, Y, and Z)
- **Gyroscopes**: These measure the angular velocity or the rate of rotation around different axes
- **Magnetometers** (optional): These measure the strength and direction of the magnetic field

An IMU consists of the sensors mentioned in the following list, which are primarily present in the flight controller, meaning they are helpful in determining vehicle position and orientation. However, some IMUs come with the magnetometer built-in, and for some devices, we need to add this externally. The following are the main sensors used in the IMU:

- **Accelerometer**: This measures the acceleration along its axis.
- **Gyroscope**: This measures the angular velocity and acceleration.
- **Magnetometer**: This is an optional sensor in an IMU; some IMUs do come with the magnetometer, and some do not. A flight controller mostly comes with an accelerometer and gyroscope; however, a barometer and magnetometer are optional.

For proper flight, a drone requires IMU. The drone's internal sense of balance, or IMU, keeps it level and facilitates smooth flight. The IMU provides data to the drone's computer, which utilizes it to determine its motion. It functions similarly to the drone's means of ensuring that it maintains control and complies with the user's instructions. The data processing of the IMU enables the drone to fly precisely and perform as intended.

Accelerometer

An accelerometer measures acceleration, that is, the rate of a change in speed, to understand the changes in speed and direction. It helps in stability, helping flight controllers maintain flight levels and respond to pilot commands. The data from all three axes of an accelerometer are processed by the flight controller to compute the rate of change in the speed of the drone.

By measuring acceleration along each axis, the flight controller can integrate these data over time to calculate changes in velocity. Essentially, the accelerometer provides information about the drone's acceleration or deceleration rate in different directions, allowing the flight controller to make real-time adjustments to maintain stability.

Examples of MEMS accelerometer devices used in flight controllers:

- STMicroelectronics LSM6DS3
- InvenSense MPU-6050
- Bosch Sensortec BMI160
- ADXL345

The following is an image of a MEMS accelerometer:

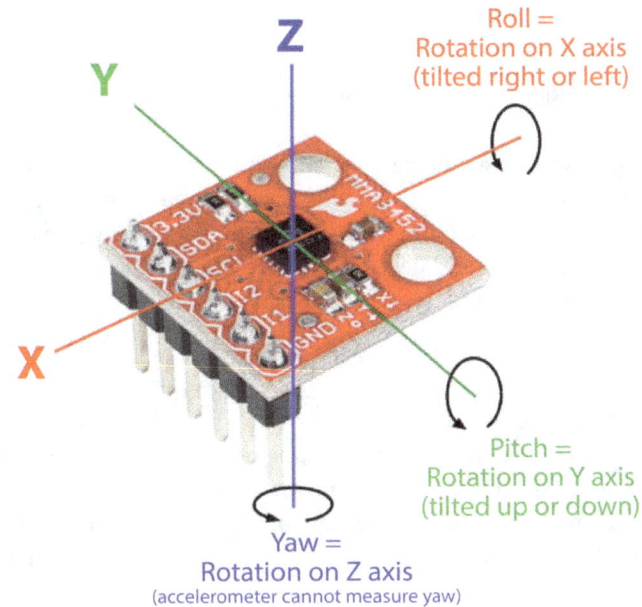

Figure 5.1 – MEMS Accelerometer

A standard accelerometer measures the force of gravity on it across an axis. A standard accelerometer on Earth (at rest) would measure an acceleration of 9.8 m/sec^2. When the system is in motion, it gives the acceleration data, which helps derive the speed, movement, and position of the drone.

Gyroscope

A gyroscope sensor measures the rate of rotation or angular velocity around different axes. In a drone, gyroscopes help to determine the drone's orientation and provide stability during flight. The data from the gyroscope helps the flight controller to understand how and in which direction the drone is turning and allows it to make real-time adjustments to maintain a level flight by adjusting the motor speed. For example, when a drone tilts or rotates to one side, the gyroscope data informs the flight controller, who then adjusts the motor speed on that side to counteract the movement and keep the drone stable and leveled:

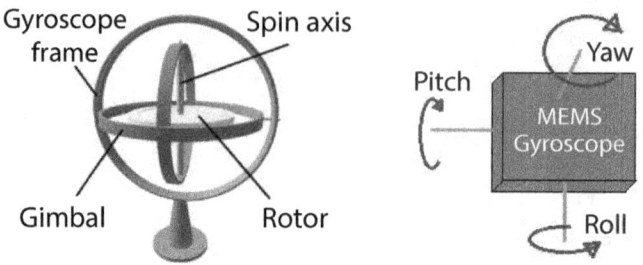

Figure 5.2 – Mechanical three-axis gyroscope

An example of a MEMS gyroscope used in flight controllers nowadays is the InvenSense MPU-6050. It combines a three-axis gyroscope with a three-axis accelerometer. The integration of gyroscope data, along with accelerometer and, sometimes, magnetometer data, provides a comprehensive set of information about the drone's motion. This integrated data is crucial for the flight controller to accurately determine the drone's orientation in three-dimensional space, enabling stable and controlled flight.

Magnetometer

A magnetometer sensor provides the measurement of the intensity and orientation of the magnetic field of the Earth. In drones, it is used to determine the heading or orientation of the drone with respect to Earth's magnetic field or magnetic north. The heading information helps in navigation and is critical to the drone's autonomous movement to defined waypoints. Nowadays, a MEMS magnetometer is merged into the IMU together with accelerometers and gyroscopes. However, certain flight controllers tend to keep it separated.

An example of a widely used magnetometer is Bosch Sensortec BMM150, which is a MEMS magnetometer usually found in drone flight controllers. The magnetometer data are merged with other sensor inputs using sensor fusion algorithms; this helps to compute the exact position and orientation of the drone, contributing to fine control over its navigation while flying in autonomous missions.

The following is a diagram that shows the Earth's magnetic field and its magnetic heading and inclinations:

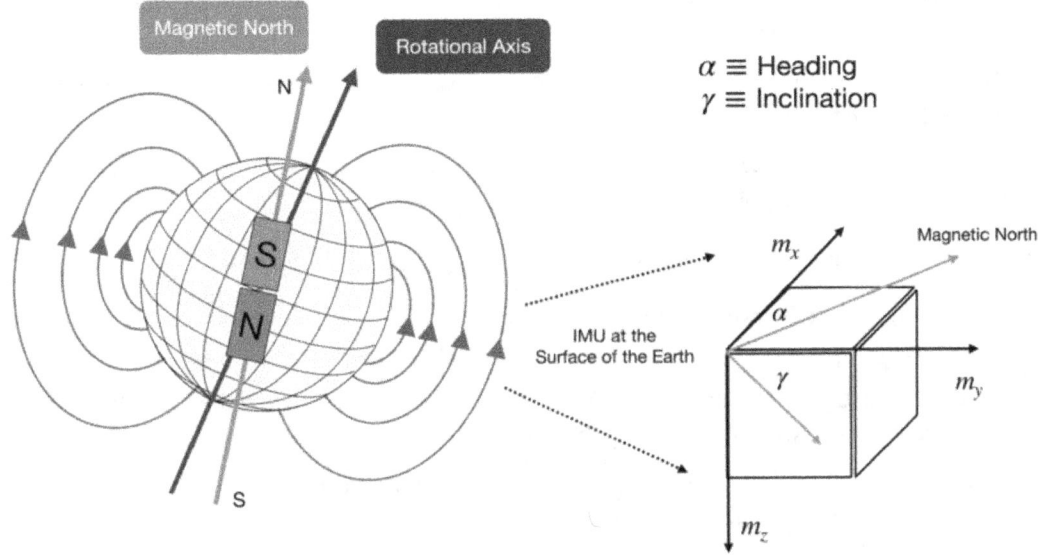

Figure 5.3 – Earth's magnetic field

Use of IMUs in unmanned systems

Nowadays, the number of IMUs can vary based on the complexity of the drone and its application. Many drones have a single IMU, while many others designed for professional applications have redundant IMUs for reliability and accuracy.

Redundant IMU setup

This is found in flight controllers and is intended to be used for high-end professional drones, such as industrial drones, and those designed for mission-critical applications. The prime reason for redundant IMUs is that they serve as a backup in case one unit fails, and a failover can switch to secondary IMUs, enhancing the overall reliability. Let's see what a triple redundant IMU setup consists of the following:

- **Configuration**: There are three independent IMUs that are integrated into the drone's flight controller across the three axes
- **Redundancy**: Each IMU consists of accelerometers, gyroscopes, and, sometimes, magnetometers, providing redundant data on the drone's motion and orientation
- **Algorithm**: Modern flight controllers deploy sensor fusion algorithms that can process data from all three IMUs to determine the most accurate and reliable information

The prime features of a redundant IMU system are the following:

- **Failover mechanism**: This provides a fail-safe mechanism. If one IMU malfunctions, the system seamlessly switches to the backup
- **Accuracy**: Redundant IMUs can be used for cross-validation, improving overall accuracy in determining the drone's orientation and motion
- **Mission-critical applications**: Ideal for missions where reliability and precision are crucial, such as surveying, mapping, or inspections

The use of an IMU in an advanced flight controller:

- **Positioning and navigation**: A single or a redundant IMU plays an important role in determining the drone's position and its orientation, aiding accurate navigation
- **Autonomous flight**: IMU data aid in autonomous flight modes, allowing the drone to navigate and perform tasks without constant pilot input
- **Safety features**: In combination with other sensors, IMUs contribute to safety features such as emergency landing, obstacle avoidance, and geofencing

The barometer and its role

Barometers provide height information that helps in the accurate control and navigation of a drone. Basically, atmospheric pressure is measured by barometers since air pressure decreases as the altitude increases, and this informs the flight controller of how high it has ascended.

With regard to altitude control, the drone flight controller uses barometer data to estimate the present altitude above sea level. This allows the flight controller to calculate the vertical position of a drone precisely.

The barometer data are fused with other sensors, such as accelerometers and gyroscopes. Thus, it is possible to observe the sensor values in the IMU part of the flight controller. The following two barometers—Bosch Sensortec BMP280 and BMP388—are mostly used in drone flight controllers; they are known to establish co-operation with the IMU that enables a more thorough measurement of data concerning the control of altitude, navigation, and stability.

A barometer interface with I2C or SPI sends the altitude signal data to the flight controller. The flight controller then sees the required altitude and pushes the motors to produce more thrust and, hence, gain the desired altitude. The desired altitude will be shown by the barometer sensor.

How does it work?

In a flight controller, a barometer estimates altitude by measuring the change in atmospheric pressure as the drone ascends or descends. The basic principle behind this is that the air pressure decreases with increasing altitude. The barometer provides the current atmospheric pressure data to the flight controller at the drone's current location. A barometer needs to be calibrated to a reference pressure point first to provide the reference data to the flight controller. At sea level, the flight controller calculates the altitude above that calibrated reference point. Different height references include the following:

- **Above-ground level (AGL)**: The AGL represents the height above the ground or terrain at that moment, providing a measure relative to the Earth's surface
- **Above mean sea level (AMSL)**: The AMSL represents the height above a standard reference point, normally sea level, offering a consistent elevation measure regardless of local terrain variations.

The flight controller uses this barometric altitude information for different purposes, such as stable flight, executing precise maneuvers, and navigating in three-dimensional space.

GPS and magnetometer and their roles

The GPS and magnetometer are the two main and interrelated components of the drone. Both of the components work as the eyes and ears of the drone to help the drone locate itself in the 3D environment with respect to the Earth and also to navigate in the 3D space. The presence of a GPS and compass helps the drone navigate through different waypoints in space completely autonomously and also while returning to the launch position:

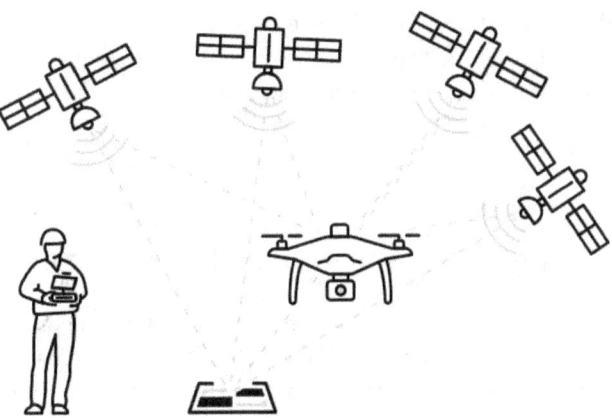

Figure 5.4 – An image depicting how a GPS works

How does GPS work?

The GPS module is used to interface with the flight controller by any means of a hardware protocol, i.e., SPI, I2C, or CAN, depending on the manufacturer and model (see product user manual). A GPS module consists of different positioning systems, such as GPS, Glonass, Beidou, and so on, which receive the signals from the satellite network via radio receivers. These signals have the velocity, time, and location co-ordinate data that feed the flight controller. The flight controller uses these co-ordinates to place itself in relation to Earth, and uses with these co-ordinates when estimating velocity, position, and time. These data, again, are sent to the sensor fusion algorithm in the flight controller and are fused with the IMU data to give the correct estimates.

The use of GPS in drones

Though drones can be flown manually without GPS, GPS is required for the drone to function in auto mode and conduct waypoint missions. The following are the benefits of GPS:

- **Stability**: The stability of the flight is increased by installing a GPS. The GPS has fixed co-ordinates regarding the position of the drone. Hence, it helps the drone not drift from its original position by any sources, such as wind, and stabilizes its position using the co-ordinates; however, it depends upon the quality of the GPS and its accuracy. The better the quality of the GPS, the better it can hold the position.
- **Return to home**: During the emergency recovery procedure, this is the best feature to safeguard the drone from flying away and crashes. The original GPS co-ordinates are saved in the drone, and in case of any mishap or loss of signals, it automatically comes back and lands at the original co-ordinates. This helps the safe recovery of the drone.
- **Navigation**: Navigation is an important aspect of autonomous missions. Every time an autonomous mission is planned, a few GPS co-ordinates are given for the drone to automatically follow. Due to the help of GPS, a smooth auto flight is governed, which is primarily used in the commercial application of surveys, surveillance, and agriculture.

The role of GPS is quite wide—we have covered only a few aspects of it here. We will see different GPS applications in the coming chapters.

Voltage and current sensors and their roles

Current and voltage measurements in drones are essential for monitoring the electrical system's health and power management of the drone, particularly the consumption and battery status. Let's see how voltage and currents are measured using different onboard sensors:

- **Current measurement**:
  - **Sensors**: Generally, a Hall-effect current sensor (for example, ACS712) is used to measure the voltage and current flow.
  - **Working**: Hall-effect sensors detect the amount of magnetic field generated by the flow of current through a conductor. The sensor outputs an analog voltage proportional to the current strength, which is then read by the flight controller to make respective decisions.

- **Interface**: Analog voltage is usually fed into an **analog-to-digital converter** (**ADC**) on the flight controller, providing a digital representation of the current.

• **Voltage measurement**:

- **Sensor**: Voltage divider or voltage measurement IC (for example, INA219, INA226, etc.)
- **Working**: Voltage divider ICs use resistors to scale down the battery voltage to a level that is suitable for measurement. Voltage Measurement ICs directly measure the voltage and provide a digital output. These digital outputs are a modern way to measure the current and voltage.
- **Interface**: The output from the voltage divider or IC is connected to an ADC on the flight controller, converting the analog signal into a digital value for processing.

The use of voltage and current measurements:

- **Low-voltage protection**: The flight controller uses voltage measurements to trigger safety warnings or initiate safety procedures when the battery voltage is critically low. Typically, these values are set by the pilot before flight and trigger safety behavior such as RTL or land.
- **Overcurrent protection**: Current measurements help detect overcurrent situations, allowing the flight controller to implement protective measures.

Sensor fusion and state estimation

Sensor fusion and state estimation represent an important part of flight controller algorithms, as this is the function that helps the drone combine all the data coming into it via the IMU, compass, and GPS; the system then makes a decision on the exact value to velocity, speed, altitude, direction, and angular acceleration. This data is important, as it helps the flight controller to make calculated decisions in terms of maneuverability:

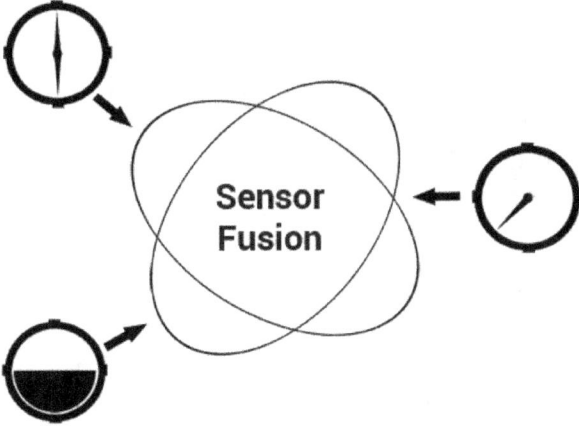

Figure 5.5 – The concept of sensor fusion

The state estimation algorithms take data from the accelerometer and GPS co-ordinates, fuse them together, conduct a sanity check, and pass them to the flight controller, who observes the current acceleration and speed of the drone. Sensor fusion also takes care of errors in the reading of some sensors and ignores them in the decision-making process.

Summary

In this chapter, we have gone through the different sensors that are present in a drone and how each sensor is useful for stable and safe flight. Knowledge of these sensors helps the system engineer design new applications, and this also helps firmware engineers design new algorithms to play with them.

In the next chapter, we will learn more about how these sensors help in flying in different flight modes and how these data are interpreted by the flight Controller.

6
Introduction to Drone Firmware and Flight Stacks

In previous chapters, we took a look at major hardware and peripheral devices, as well as various sensors that are helpful for the control, command, guidance, and navigation of a drone. These are the mechanical as well as electronic parts that are interfaced and assembled to form the system hardware that we call a drone. But you will know that hardware is nothing without a soul, and the soul of the drone is the drone firmware. To control or command the drone as a whole, we require software called **Ground Control Station** (**GCS**) software, which we will study briefly in the next chapter.

In this chapter, we will go through the high-level architecture of the firmware, what it requires, what it includes, and the software that interfaces with it to form the drone. We will go through the following topics, which will be helpful for us to better understand what a firmware consists of and other helpful libraries that we will be making use of.

The following are the main topics covered in this chapter:

- Introduction to firmware
- Introduction to drone firmware
- Introduction to drone flight stacks
- The PX4 flight stack
- The ArduPilot flight stack

This chapter will focus more on the general aspects and structure of the code base rather than the complete code base. We will not cover the entire code base and how it was developed since that would require a whole other chapter. Rather, we will look at what the essential parts are and the complete stack as a whole.

Technical requirements

In this chapter, we will focus on the architecture and software module. Knowledge of a programming language such as C++ or Python, a development environment such as **Visual Studio** (**VS**) Code or Eclipse, and a version control system such as Git and GitHub would be helpful.

Introduction to firmware

Firmware is a special type of software that is specifically designed to be installed on embedded hardware. The prime responsibility of the firmware is to control and manage the functionality of a device's specific hardware components. Firmware is stored in **Non-Volatile Memory** (**NVM**) and is persistent, meaning it retains its programming even when the power is turned off. It acts as a bridge between the hardware and higher-level software, facilitating communication and ensuring the proper operation of the device.

Firmware comes in various types, each tailored to the specific needs of the devices or systems it controls. Here are different types of firmware along with examples:

- **BIOS/UEFI firmware**:

    Description: **Basic Input/Output System** (**BIOS**) or **Unified Extensible Firmware Interface** (**UEFI**) firmware initializes hardware components during the bootup process of a computer.

    Example: UEFI firmware in modern PCs.

- **Embedded firmware**:

    Description: Used in embedded systems, such as IoT devices, consumer electronics, and industrial machines, to control specific hardware functions.

    Example: Firmware in a smart thermostat that manages temperature control and user interfaces.

- **Bootloader firmware**:

    Description: Responsible for loading the **Operating System** (**OS**) into memory during the boot process, performing system checks, and providing basic configurations.

    Example: **Grand Unified Bootloader** (**GRUB**) in Linux systems.

- **Peripheral firmware**:

    Description: Controls the operation of hardware peripherals such as printers, cameras, network devices, and other external devices.

    Example: Firmware in a printer that manages print jobs, paper handling, and communication with the computer.

- **Microcontroller firmware**:

    Description: Used in microcontrollers to manage input/output operations, interact with sensors and actuators, and execute specific tasks.

    Example: Firmware in an Arduino microcontroller that controls attached sensors, LEDs, or motors.

- **Network device firmware**:

    Description: Manages network protocols, security features, and communication with other devices in routers, modems, and network switches.

    Example: Firmware in a Wi-Fi router that controls network configurations and security settings.

- **Storage device firmware**:

    Description: Found in storage devices such as hard drives and **Solid-State Drives (SSDs)**, managing data storage, retrieval, and error correction.

    Example: Firmware in an SSD that optimizes data placement and wear-leveling algorithms.

- **Automotive firmware**:

    Description: Controls various systems in vehicles, including **Engine Control Units (ECUs)**, infotainment systems, and safety features.

    Example: Firmware in a car's ECU that manages fuel injection, ignition timing, and emissions control.

- **GPU firmware**:

    Description: Found in **Graphics Processing Units (GPUs)**, it manages graphics rendering, shader execution, and other GPU-specific functionalities.

    Example: Firmware in an NVIDIA or AMD GPU that optimizes graphics processing.

- **Camera firmware**:

    Description: Controls the operation of digital cameras, managing image processing, focus, exposure, and other camera-specific functions.

    Example: Firmware in a DSLR camera that optimizes image capture settings.

Components of firmware and their roles

The components of firmware play critical roles in controlling and managing the functionality of a device's specific hardware. These components work together to ensure the proper operation of the device. Here are key firmware components and their roles:

- **Bootloader**: The bootloader is the initial firmware that runs when the device is powered on. It performs hardware initialization and system checks and loads the OS or main firmware into memory.

- **Kernel or OS**: In some cases, firmware includes a simplified OS or kernel. This component manages system resources and provides a platform for higher-level applications.
- **Device drivers**: Device drivers facilitate communication between the OS or application software and specific hardware components. They enable the control and utilization of features such as sensors, peripherals, or communication interfaces.
- **Application code**: This is the main program or set of programs that define the device's primary functionality. It handles specific tasks such as data processing, user interface control, and interactions with external devices.
- **Configuration settings**: Firmware often includes settings and configurations that can be customized to adapt the device to different environments or user preferences. These settings may affect aspects such as performance, power consumption, or communication parameters.
- **Security features**: Many firmware implementations incorporate security features to protect against unauthorized access, data breaches, or other potential threats. Security measures may include encryption, access controls, and secure boot mechanisms.
- **Boot configuration**: Firmware manages the boot configuration, specifying the order and source of bootable devices. This includes settings such as boot sequence, boot device priorities, and boot timeout values.
- **Update mechanisms**: Firmware includes mechanisms for updating its code to address bugs, improve performance, and add new features. These update processes may involve **Firmware-over-the-Air** (**FOTA**) updates or flashing new firmware via physical connections.
- **Error handling and logging**: Firmware includes mechanisms for detecting and handling errors that may occur during operation. It may log error information for debugging purposes or provide feedback to users about the system status.
- **Interrupt Service Routines (ISRs)**: In real-time systems, firmware includes ISRs that handle interrupts generated by hardware events. These routines ensure a timely response to critical events and help maintain system stability.
- **Power management**: For devices with power constraints, firmware includes power management features to optimize energy usage. This may involve controlling sleep modes, adjusting clock frequencies, or selectively activating/deactivating components.
- **Memory management**: Firmware manages memory efficiently to minimize resource usage and prevent issues such as memory leaks or fragmentation. It allocates and deallocates memory as needed for different tasks.

Firmware example

Let's consider a simple example of firmware code written in C for a hypothetical embedded system controlling an LED. This example uses a popular microcontroller, the Arduino, and its development environment:

```c
// LED Blinking Firmware for Arduino

// Define the pin connected to the LED
#define LED_PIN 13

void setup() {
// Initialize the LED pin as an output
pinMode(LED_PIN, OUTPUT);
}

void loop() {
  // Turn the LED on
  digitalWrite(LED_PIN, HIGH);
  delay(1000); // Wait for 1 second

  // Turn the LED off
  digitalWrite(LED_PIN, LOW);
  delay(1000); // Wait for 1 second
}
```

In this example, the firmware code uses the Arduino framework to control the state of an LED. The `setup` function initializes the pin connected to the LED as an output, and the `loop` function repeatedly turns the LED on and off at one-second intervals.

Tools used to develop firmware

The following are the tools used to develop firmware:

- **Integrated Development Environment (IDE)**:

    Role: The IDE provides a platform for writing, compiling, and uploading firmware code to the target microcontroller.

    Examples: Arduino IDE, MPLAB X, Keil µVision, and PlatformIO.

- **Compiler**:

    Role: The compiler translates the high-level firmware code written by developers into machine-readable instructions for the specific microcontroller.

    Examples: GCC, Keil, and MPLAB XC.

- **Debugger**:

    Role: The debugger facilitates debugging and testing of the firmware by allowing developers to inspect variables, set breakpoints, and step through code.

    Examples: In-circuit debugger and JTAG debugger.

- **Programmer/debugger hardware**:

    Role: The programmer/debugger hardware connects the development environment to the target microcontroller, allowing code uploading, debugging, and in-circuit programming.

    Examples: JTAG programmer, USBasp, and ST-LINK.

- **Microcontroller development board**:

    Role: The microcontroller development board provides a platform with the necessary hardware components (microcontroller, input/output pins, and power supply) for testing and running firmware.

    Examples: Arduino boards, STM32 Nucleo boards, and PIC development boards.

- **Oscilloscope/logic analyzer**:

    Role: The oscilloscope/logic analyzer helps analyze and debug the electrical signals at various points in the circuit, which is useful for optimizing timing and diagnosing issues.

    Examples: Saleae Logic and Rigol Oscilloscopes.

- **Text editor/IDE extensions**:

    Role: Developers often use text editors or IDE extensions to enhance their coding experience with features such as syntax highlighting and code completion.

    Examples: VS Code and Sublime Text.

- **Version control systems**:

    Role: Version control systems manage versioning and collaborative development, allowing multiple developers to work on the firmware code simultaneously.

    Examples: Git and SVN.

- **Simulation and emulation tools**:

    Role: Simulation and emulation tools simulate the behavior of the firmware on a virtual platform, aiding in testing and debugging before deploying on physical hardware.

    Example: Proteus and QEMU.

- **Documentation tools**:

    Role: Documentation tools generate documentation for the firmware code, making it easier for developers to understand and maintain the code base.

    Example: Doxygen and Javadoc.

We have now seen examples of firmware and development tools. These would be the same for any kind of software and firmware you want to develop and install.

Differences between software and firmware

Firmware and software represent distinct categories of computer programs with key differences in their scope and functionality. Firmware is a specialized type of software closely tied to the hardware of embedded devices. It resides in NVM, providing low-level control over specific hardware components. Notably, firmware persists even when the device is powered off and is responsible for device initialization and management. In contrast, software encompasses a broader range of programs that operate within a computer system, interacting with the hardware through the OS. Unlike firmware, software typically loads into volatile memory and does not persist across power cycles. Software offers extensive customization options, is often updated more easily, and is developed with a focus on higher-level programming languages. In summary, firmware is intimately connected to hardware, providing foundational control, while software operates at a higher level, delivering diverse applications and functionalities within a computing environment.

Here's a comparison between firmware and software at the code and architectural levels presented in a tabular format:

Aspect	Software	Firmware
Execution Environment	Typically runs on general-purpose processors	Runs on dedicated processors or microcontrollers
Programming Language	Higher-level languages (C++, Java, Python, etc.)	Lower-level languages (C and assembly)
Memory Usage	Utilizes RAM and other volatile memory extensively	Designed to minimize RAM usage; often uses Flash memory
Real-Time Constraints	May not have strict real-time constraints	Often requires adherence to real-time constraints
Error Handling	Tends to rely on exceptions and error-checking mechanisms	May use simpler error-handling mechanisms due to resource constraints
Hardware Interaction	Primarily relies on OS APIs for hardware access	Directly interacts with specific hardware components and peripherals
Abstraction Level	Operates at a higher level of abstraction	Operates at a lower level, closer to the hardware
Resource Optimization	Emphasizes optimization for performance and resource efficiency	Prioritizes resource efficiency given hardware constraints

Development Environment	Developed using general-purpose development tools and environments	Often requires specialized development tools and hardware-specific considerations
Portability	More portable across different hardware platforms	Specific to the hardware architecture and may require modification for portability
Example Code Snippet	python print("Hello, World!")	c int main() { printf("Hello, World!"); return 0; }

Table 6.1 – Comparison of software and firmware

We have now seen the difference between software and firmware, their roles, and what tools are required for development. In the following section, we will go ahead and explore the firmware used in a drone.

Introduction to drone firmware

Drone firmware is specialized software that is designed to control the flight, navigation, and overall operation of **Unmanned Aerial Vehicles (UAVs)**, commonly known as drones. It plays an important role in managing the drone's hardware components, processing sensor data, executing control algorithms, and ensuring stable and responsive flight. Drone firmware is unique in that it addresses the specific challenges and requirements of aerial vehicles, providing control over various aspects, such as motor outputs, attitude, altitude, and position.

How it is different from general firmware

While general firmware refers to software that controls the operation of electronic devices, drone firmware is tailored to the specific needs of unmanned aerial systems.

The key differences include the following:

- **Control complexity**: Drone firmware incorporates sophisticated control algorithms to manage flight dynamics, stability, and navigation, which are not typically found in general-purpose firmware.
- **Sensor integration**: Drones rely on a variety of sensors (accelerometers, gyroscopes, barometers, and GPS) to maintain stable flight. Drone firmware is designed to integrate and process data from these sensors for accurate control.
- **Navigation and autonomy**: Unlike a lot of general-purpose firmware, drone firmware often includes features for autonomous flight, waypoint navigation, and obstacle avoidance.
- **Real-time operation**: Drones require real-time responsiveness to maintain stability during flight. Drone firmware is optimized for low-latency operation, ensuring quick and precise adjustments.

The structure of drone firmware

The structure of drone firmware is organized to handle different aspects of flight control, sensor integration, communication, and user customization.

The firmware architecture in a multi-copter is a complex system designed to control and manage various aspects of the aircraft's flight. Here is a detailed explanation of the components and their roles in the firmware architecture:

- **Main control loop**:
  - **Role**: The central control loop that governs the overall flight behavior of the multicopter
  - **Components**:
    - **Sensors**: Collect data such as accelerations, gyroscopic rates, and altitude
    - **Flight controller**: Processes sensor data, executes control algorithms, and generates motor commands
    - **Actuators (motors)**: Respond to motor commands, influencing the multicopter's movements

- **Sensor integration and fusion**:
  - **Role**: Collects and combines data from multiple sensors to create a comprehensive understanding of the multicopter's state
  - **Components**:
    - **Inertial Measurement Unit (IMU)**: Combines accelerometer and gyroscope data to determine orientation
    - **Barometer**: Measures atmospheric pressure for altitude estimation
    - **GPS receiver**: Provides global position data for navigation and position holding

- **Attitude controller**:
  - **Role**: Maintains the multicopter's desired orientation (roll, pitch, and yaw)
  - **Components**:
    - **Proportional-Integral-Derivative (PID) controller**: Adjusts motor outputs based on the difference between desired and actual attitude

- **Altitude controller**:
  - **Role**: Manages the multicopter's altitude during flight
  - **Components**:
    - **PID controller**: Regulates motor outputs to control the multicopter's vertical position
- **Position controller**:
  - **Role**: Controls the multicopter's position in space, particularly in horizontal directions
  - **Components**:
    - **PID controller**: Adjusts motor outputs based on the difference between the desired and actual position
- **Navigation and waypoint handling**:
  - **Role**: Enables autonomous flight by processing GPS data and following predefined waypoints
  - **Components**:
    - **Navigation algorithms**: Determine the optimal path based on GPS coordinates
    - **Waypoint manager**: Coordinates the multicopter's movements to follow a predefined route
- **Communication interface**:
  - **Role**: Facilitates communication between the multicopter and external devices, such as GCSs or remote controllers
  - **Components**:
    - **Telemetry system**: Sends real-time flight data to ground control
    - **Radio receiver**: Receives user input from the remote control
- **Power management**:
  - **Role**: Monitors and manages the power system, including battery voltage and current
  - **Components**:
    - **Battery monitor**: Ensures safe battery usage and provides low-battery warnings
- **Motor and ESC interface**:
  - **Role**: Translates motor commands generated by the flight controller into specific signals for the **Electronic Speed Controllers (ESCs)** and motors

- **Components**:
    - **Pulse-Width Modulation (PWM) generator**: Generates signals for ESCs based on control inputs
- **Configuration and settings**:
    - **Role**: Allows users to customize parameters and settings for the multicopter's behavior
    - **Components**:
        - **Configuration storage**: Saves user-defined settings for future flights
- **Flight logging**:
    - **Role**: Records flight data for analysis and troubleshooting
    - **Components**:
        - **Data logger**: Captures sensor readings, control inputs, and other parameters during flight

This firmware architecture operates in a continuous loop, with the main control loop orchestrating interactions between various components. The integration of sensors, controllers, and communication interfaces ensures that the multi-copter can achieve stable flight, respond to user inputs, and execute autonomous tasks when required. Advanced firmware architectures may include additional components for obstacle avoidance, computer vision, or other specialized functionalities.

Introduction to drone flight stacks

A flight stack is a collection of software components that control the flight of a drone. So, we can say that an open source flight stack is one where the source code of the software components is made available to the public to use, view, modify, and distribute among users, to enhance public collaboration and joint development. On the other hand, a proprietary drone flight stack is one that is developed by a particular company or entity that owns it and its source code is not publicly accessible to view, modify, and distribute.

Next, we will look at the main differences between an open source and closed source flight stack for better understanding.

Open source drone flight stack

Source code: The source code in an open source drone flight stack is made freely available to the public, which allows anyone to view, modify, and distribute the code.

Collaboration: A wide range of community developers contribute to open source projects that encourage community collaboration, leading to fast development, identified bug fixes, and the development of new features.

Conceptual understanding: Access to the source code of open source projects helps users understand how the software stack works together. Users can scrutinize the code, which leads to increased trust and security.

Modification: Users have the comfort of amending the source code to fulfill their specific needs or develop new features. This flexibility is beneficial for researchers, hobbyists, and organizations with unique experiments and tests.

Price: Open source software is free to use. Users can download, use, and modify the software without incurring licensing fees.

The following are some examples of open source flight stacks:

- **PX4 Autopilot**: PX4 is an open source project maintained by the Dronecode project, with contributions from a global community of developers
- **ArduPilot**: ArduPilot is an open source project with contributions from a diverse community of developers worldwide
- **Paparazzi UAV**: Paparazzi UAV is an open source project developed by a community of researchers, students, and hobbyists
- **DroneKit**: DroneKit is an open source project developed by **3D Robotics (3DR)** and the community

Closed source (proprietary) drone flight stack

Source code availability: The source code is not publicly available in a closed source drone flight stack. It is owned by a specific company or entity and is proprietary, providing intellectual property protection.

Control: The company that owns the closed source stack has complete control over the development, updates, and feature releases. Users depend on the company for support and updates.

Limited collaboration: There are limited community collaborations but major development is executed by the internal development team of the company that owns the proprietary stack. External contributions from the broader community are not typically accepted.

Support: The user's dependency on the company increases for users of a closed source solution. Updates and new features are provided according to the company's schedule.

Price: There are licensing fees for closed source solutions; users may need to pay for the software and any additional features, and ongoing support services may incur additional costs.

Some examples of a closed source flight stack are as follows:

- **DJI Flight Stack (DJI SDK)**: DJI, a leading drone manufacturer, develops proprietary flight stacks for their drones. The DJI **Software Development Kit** (**SDK**) is available for developers to build applications on top of DJI's flight control systems.
- **Yuneec flight stack**: Yuneec International is a drone manufacturer that develops proprietary flight stacks for their drones.
- **Autel Robotics flight stack**: Autel Robotics, another drone manufacturer, develops proprietary flight stacks for their drones.

To have an easy-access view of the features of the different flight stacks, here is a tabular comparison outlining the major technical differences between ArduPilot, PX4, INAV, DroneKit, and Paparazzi:

Feature/Aspect	ArduPilot	PX4	INAV	DroneKit	Paparazzi
Architecture	Mix of C++ and a scripting language (APL)	C/C++ with NuttX real-time OS	Primarily C, designed for fixed-wing	Python SDK, communicates with MAVLink	Mix of C and Python scripts
Middleware	Robot Operating System (ROS)	micro Object Request Broker (uORB)	No specific middleware, relies on standard libraries	MAVLink for communication	Custom middleware
Communication Protocol	MAVLink	MAVLink	MAVLink	MAVLink	Custom (UDP, serial, or Ethernet)
Primary Language	C++	C/C++	C	Python	C or Python
License	GPLv3	BSD 3-clause license	GPLv3	Apache 2.0	GPLv2
Real-Time Support	Limited	Yes	No	No	Limited
Mission Planning Tools	Mission Planner, APM Planner	QGround Control, MAVSDK	INAV Configurator	Custom scripts, QGroundControl support	Paparazzi ground segment
API Availability	Rich API for customization and integration	Extensive APIs for customization and integration	Limited APIs for custom scripts and modules	DroneKit APIs for Python scripting	Limited APIs for customization

Flight Modes	Versatile, supports various flight modes	Extensive list of flight modes	Focused on GPS-assisted navigation and stabilization	Depends on APIs used for flight control	Limited modes, customizable with scripts
Community and Support	Large, active community	Active community	Active community	Moderate community	Active community
Documentation Quality	Extensive documentation, user-friendly	Comprehensive documentation, developer-friendly	Good documentation, user-friendly	Moderate documentation	Limited documentation, steep learning curve
Autonomous Features	Rich set of autonomous features and missions	Feature-rich with a focus on versatility and modularity	Primarily designed for navigation, less focus on autonomous missions	Provides APIs for drone control and telemetry	Basic autonomy features, emphasis on fixed-wing UAVs
Ease of Use	User-friendly interface, extensive Mission Planner	User-friendly interface, QGround Control	User-friendly interface, relatively simple configuration	Requires programming skills and Python knowledge	Complex setup, steep learning curve
Hardware Compatibility	Broad hardware support for various autopilot boards	Broad hardware support for various autopilot boards	Limited to a specific set of boards	Compatible with various drones and hardware	Limited hardware compatibility
Customization Options	Highly customizable with extensive parameters	Highly customizable, modularity is a key feature	Limited customization options	Customizable through Python scripts	Customizable through C and XML configuration files
Ground Control Software	Mission Planner, APM Planner	QGround Control, MAVSDK	INAV Configurator	Various ground control software supporting MAVLink	Paparazzi ground segment

Table 6.2 – Comparison of different flight stacks

In the next few subsections, we will learn about drone firmware and the firmware stacks that are used extensively in drones nowadays. Some firmware stacks are open source, which means they are free to use and can be modified by anyone, while others are closed source – that is, they are proprietary to the creator of the company and cannot be used freely by the general public. Let's go ahead and look at these.

Basic PX4 controller loop diagram

The following is a loop diagram of a drone using the PX4 autopilot stack that shows the interlinkage between different software and hardware components of the drone based on which the drone works.

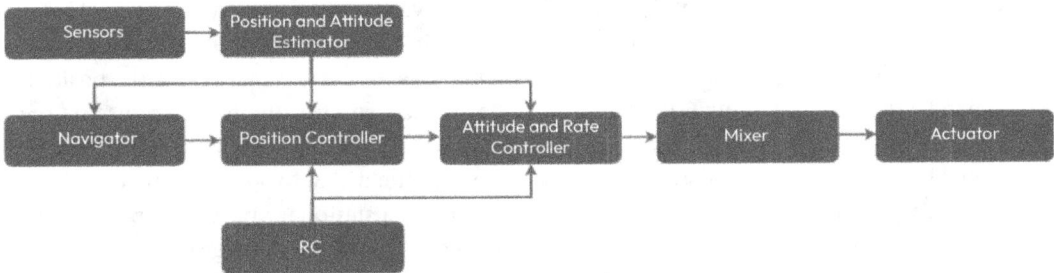

Figure 6.1 – A PX4 loop diagram

We will look at the different stages in detail, as follows:

- **Sensors**: As we discussed earlier, sensors are embedded into flight controllers and peripheral devices, such as a GPS compass, which helps the drone to get its GPS positions as well as its speed, linear velocity, and angular velocity, which is guided by position and attitude estimators.

- **Position and attitude estimators**: The position and attitude estimators in PX4, specifically the **Extended Kalman Filter 2** (**EKF2**), play a pivotal role in determining a drone's location and orientation during flight. The EKF2 position estimator utilizes data from various sensors, including GPS, accelerometers, gyroscopes, magnetometers, and barometers, to estimate the drone's 3D position (latitude, longitude, and altitude) and velocity. Simultaneously, the attitude estimator employs data from gyroscopes and accelerometers, augmented by magnetometer data for improved yaw estimation, to determine the drone's attitude in terms of roll, pitch, and yaw angles. These estimators utilize sensor fusion techniques to mitigate noise and inaccuracies, offering adaptive capabilities to adjust to changes in the drone's dynamics or sensor characteristics. Users can configure parameters to optimize the performance of these estimators based on the specific drone and sensor setup, ensuring precise and reliable flight control within the PX4 software stack.

- **Position controllers**: Position controllers in a multicopter are important components responsible for maintaining the desired location and navigating between waypoints during flight. In station-keeping, the position controller ensures the multicopter remains fixed at a specified location, actively adjusting motor outputs to counter external disturbances and maintain a stable position. For waypoint navigation, the position controller guides the multicopter along a predefined path, interpreting GPS coordinates and adjusting motor commands to navigate through a series of waypoints. By combining sensor data, GPS information, and control algorithms, position controllers enable precise station-keeping and accurate waypoint navigation, contributing to the overall stability and autonomy of the multicopter during its mission.

- **Navigator**: The navigator in the PX4 drone firmware serves as a crucial module responsible for mission planning, path following, and waypoint navigation. It interprets high-level commands, such as waypoints or geofences, and translates them into actionable directives for the position controller. By incorporating GPS data and sensor measurements, the navigator assists the position controller in precisely guiding the drone along its designated trajectory. It provides crucial information to the position controller, allowing the drone to autonomously navigate through a predefined sequence of waypoints or maintain a stationary position, ensuring the accurate and reliable execution of mission plans. The navigator acts as a bridge between mission commands and low-level control, enhancing the overall autonomy and navigation capabilities of the drone within the PX4 firmware.

- **Attitude and rate controllers**: In PX4, attitude controllers are essential components responsible for maintaining the desired orientation of a drone, encompassing roll, pitch, and yaw angles. These controllers, often implemented using PID algorithms, continuously adjust motor outputs to minimize the difference between the drone's current attitude and the desired setpoints. Rate controllers, on the other hand, focus on controlling the rotational rates of the drone around its three axes. By using data from gyroscopes, they generate commands to stabilize and control the rotational motion. For instance, if the drone experiences an unexpected roll, the attitude controller will act to correct it, while the rate controller will manage the rotational speed to ensure a smooth adjustment. These controllers work in tandem, contributing to the stability and responsiveness of the drone in various flight conditions.

- **Mixer**: Mixer is a control algorithm that takes commands from the attitude and position controllers and drives the actuators to the required speed and position.

- **Actuators**: Actuators in PX4 play a crucial role in translating control commands from the flight controller into the physical movements of the drone's motors. These components convert the electronic signals generated by the flight controller into mechanical actions, directly influencing the rotational speeds of the motors or adjusting control surfaces on fixed-wing aircraft. For example, in a quadcopter, the actuators respond to commands from the attitude and rate controllers, modifying the thrust output of each motor to achieve the desired pitch, roll, and yaw motions. The effectiveness of the actuators is pivotal in executing precise and responsive maneuvers, contributing to the overall stability and control of the drone within the PX4 firmware ecosystem.
- **RC**: RC is used for the manual control of the drone by the pilot. The use of RC starts pilot-assisted modes to fly the drone, such as altitude and position mode. This is where the pilot give commands to the drone to fly and the position controller will behave according to the commands given by the RC.

We have seen a high-level loop diagram of how sensor data is used by controllers to drive the actuators in the given scenario of manual or auto missions. It demonstrates the flow in which different controllers in the firmware, such as position, attitude, and rate controllers, work together in the loop to provide accurate and real-time information to the actuators. It runs in a feedback mechanism where the output serves as input as well and inputs are dependent on outputs.

PX4 flight stack

The PX4 flight stack is a versatile and powerful autopilot stack that can be used to control many different kinds of unmanned vehicles, including those with fixed wings and multi-rotors, and also ground vehicles such as rovers and water vehicles such as submarines. This open source firmware is compatible with numerous pieces of flight controller hardware due to its flight controller sensors and other peripheral devices, which include high-level flight mode and safety features. The complete user and developer documentation can be found on the website: `https://px4.io/`.

If we talk about the architecture of PX4, it's quite lengthy and interconnected in nature, which we can see in *Figure 6.2*. A major part of the PX4 flight controller is its controller module, which is managed by the current state of its machine. The architecture is divided into four major parts, namely, **Storage**, **Drivers**, **External Connectivity**, and **Flight Control**, which are further divided into the modules and submodules listed next.

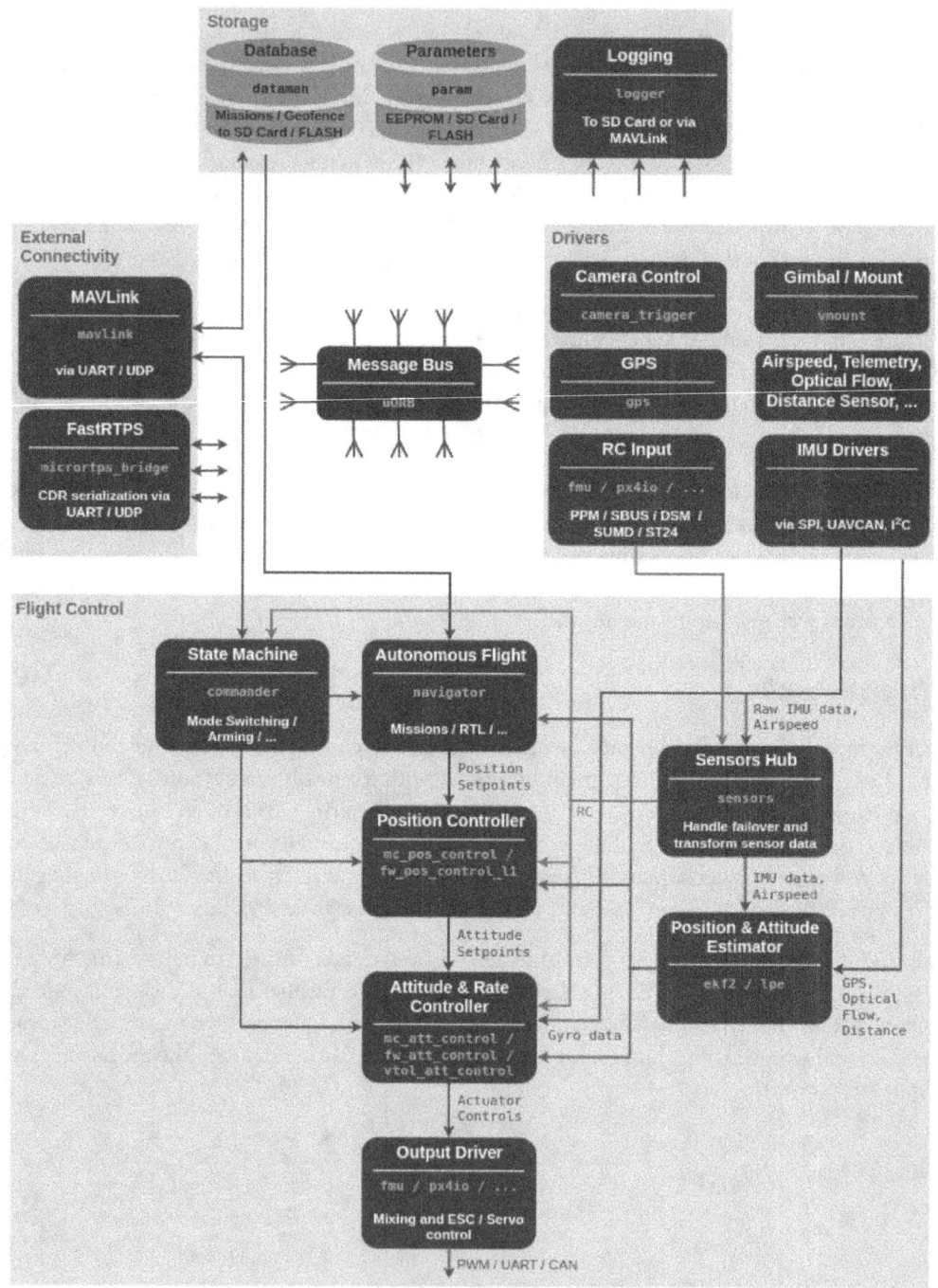

Figure 6.2 – High-level architecture diagram of the PX4 flight stack

Let us go through the different segments of the flight stack architecture to understand the role of each component:

- **Storage**: Storage is the key element of the flight stack, which is used to store the key parameters of the complete flight stack during configuration and missions. These storage mechanisms collectively ensure the availability of critical data for flight operations, firmware updates, and post-flight analysis. We can also see the interconnection between different aspects of the flight stack and their dependency on each other. The various kinds of storage available in in-flight controller are as follows:

    - **NVM**: This type of storage retains data even when power is removed. It typically stores firmware, configuration parameters, and critical flight logs.
    - **MicroSD card**: Drones often use MicroSD cards for additional storage. This can include high-resolution sensor data, detailed flight logs, and other mission-specific information.
    - **Onboard memory**: Flight controllers have onboard memory for storing firmware, configuration settings, and temporary data during flight operations.

- **Flight logs**: Flight logs contain detailed records of a drone's flight, including sensor data, GPS information, control inputs, and system states. These logs are essential for post-flight analysis, troubleshooting, and performance optimization in UAV operations.

- **External connectivity**: External connectivity refers to the two-way messages through which the flight controller interacts with external interfaces such as a GCS. External connectivity refers to the connection of the drone through the connectivity channel, such as radio telemetry via a protocol named MAVLink. This helps the GCS configure the drone and gives command and control to it, such as for planning and executing missions. The external connectivity is established via two basic protocols:

    - **MAVLink**: MAVLink is the **Micro Aerial Vehicle** protocol for unmanned systems and serves as the communication bridge between drones and GCSs. Its interface is characterized by a lightweight, binary packet structure that allows for efficient data transmission. MAVLink employs a publish-subscribe model, allowing you to subscribe to relevant messages, or publish information, facilitating seamless and modular communication. The protocol supports serial links, network connections using TCP/UDP, and radio links using the UART protocol to establish connections between drones and GCSs. We will look at this method in the GCS section when the drone is connected to the GCS.
    - **Fast RTPS**: **Fast RTPS**, or **Fast Real-Time Publish-Subscribe**, is like a high-speed communication superhero for drones. It helps drones and their peripherals talk to each other quickly and reliably. FastRTPS makes sure this communication happens in a flash, which is super important for drones to navigate and respond to commands swiftly. It connects various parts of the drone, such as the brain (flight controller) and the eyes (sensors), using several types of connections, such as wires or wireless links.

Here is a tabular representation of the technical differences between and usage of MAVLink and Fast RTPS:

Feature/Aspect	MAVLink	Fast RTPS
Communication Protocol	Lightweight messaging protocol for UAVs	General-purpose middleware for real-time systems
Usage Domain	Primarily designed for communication between UAV components and GCSs	General-purpose middleware for various real-time systems such as robotics, IoT, and industrial automation
Data Serialization	Efficient binary serialization	XML or binary serialization, customizable
Transport Layers	Supports various transport layers, including serial, UDP, and TCP	Supports transport over UDP, TCP, shared memory, and serial links
Publish-Subscribe Model	Utilizes publish-subscribe for message distribution	Follows the publish-subscribe communication paradigm, allowing multiple subscribers for a single topic
Quality of Service (QoS)	Limited QoS features, basic message reliability	Advanced QoS features, including reliability, durability, and history management
Middleware Purpose	Specialized for the UAV ecosystem, particularly in the open source drone community	General-purpose middleware, suitable for a wide range of real-time applications beyond UAVs
Community Support	Extensive community support within the drone and robotics community	General use across various industries with strong support from the Eclipse Foundation
Protocol Overhead	Lightweight with minimal protocol overhead	More comprehensive protocol with additional features, leading to slightly higher overhead
Implementation Language	Primarily implemented in C with some language bindings	Implemented in C++, with language bindings for C, Java, and others
Example Implementations	ArduPilot, PX4	Robot Operating System 2 (ROS 2), eProsima Fast RTPS

Table 6.3 – Comparison between MAVLink and Fast RTPS

While both MAVLink and Fast RTPS can be used in robotics and UAV applications, MAVLink is specifically tailored for lightweight communication between UAV components, while Fast RTPS is a general-purpose middleware with more advanced features suitable for a broader range of real-time systems.

- **Drivers**: In PX4, drivers are software components that interface with and manage the communication between the flight controller and various hardware peripherals and sensors. These drivers are crucial for the integration of different sensors and actuators into the PX4 ecosystem. They provide an abstract layer that allows the flight controller to communicate with hardware components without directly dealing with their specific details.

  The PX4 drivers are typically written by the PX4 community, which includes developers, contributors, and enthusiasts working on the open source project. They contribute to the development of drivers for a wide range of sensors, such as accelerometers, gyroscopes, magnetometers, barometers, GPS modules, cameras, and lidar.

  These drivers are used in various PX4-supported vehicles, including multirotors, fixed-wing aircraft, and other unmanned systems. Their primary purpose is to enable the flight controller to access and utilize data from different sensors and control actuators to ensure stable and controlled flight.

  Examples of PX4 drivers include the following:

  - **I2C and SPI drivers**: These drivers facilitate communication with sensors and peripherals that use the I2C and SPI protocols
  - **IMU drivers**: Responsible for interfacing with accelerometers, gyroscopes, and magnetometers to obtain orientation and motion data
  - **GPS drivers**: Enable communication with GPS modules to gather accurate position and velocity information
  - **PWM drivers**: Used to control servo motors, ESCs, and other actuators
  - **Analog-to-Digital Converter (ADC) drivers**: Interface with analog sensors to convert analog signals into digital data

- **Flight control**: In the PX4 flight control architecture, modules represent distinct software components that collaboratively manage various aspects of UAV operations. These modules work in concert to ensure a comprehensive and modular approach to flight control. Key modules include the commander, responsible for interpreting high-level commands and user inputs; the navigator, managing mission planning, waypoint navigation, and decision-making during flights; and the estimator module (EKF2), utilizing EKF to estimate the UAV's position, velocity, and attitude based on sensor data. Additionally, attitude and rate controllers play a crucial role in stabilizing the drone, while actuators translate control commands into physical movements. Sensor, mixer, and GPS modules contribute to data acquisition, controlling signal generation, and accurate navigation. The interplay between these modules defines the PX4 flight control system, ensuring a robust and flexible platform for diverse UAV applications.

In conclusion, the PX4 flight stack represents a robust and modular framework for UAV control. With a diverse array of modules, PX4 ensures precise flight control, autonomous navigation, and adaptability to various hardware setups. As an open source platform, it thrives on community contributions, making it a dynamic and continuously evolving solution for UAV enthusiasts and developers. Exploring the PX4 flight stack provides a comprehensive understanding of the intricacies behind modern drone control systems.

The ArduPilot flight stack

The ArduPilot flight stack is an open source autopilot software suite designed for controlling UAVs. Developed by a global community, ArduPilot supports a wide range of vehicles, including multirotors, fixed-wing aircraft, helicopters, ground rovers, and boats. ArduPilot encompasses both hardware and software components, providing a complete solution for drone enthusiasts, researchers, and commercial applications.

While ArduPilot shares similarities with the PX4 flight stack, there are key differences in their design philosophies and architectures. ArduPilot is characterized by a monolithic architecture, offering an integrated solution with fewer modular components than PX4. It relies on a single firmware code base that is adaptable to various vehicle types. In contrast, PX4 follows a more modular microservices-oriented architecture, allowing for greater flexibility and adaptability to different hardware and use cases. The choice between ArduPilot and PX4 often depends on specific project requirements, hardware compatibility, and user preferences. More studying of ArduPilot can be done at www.ardupilot.org.

ArduPilot flight stack architecture

A block diagram of the high-level architecture is given in the following diagram.

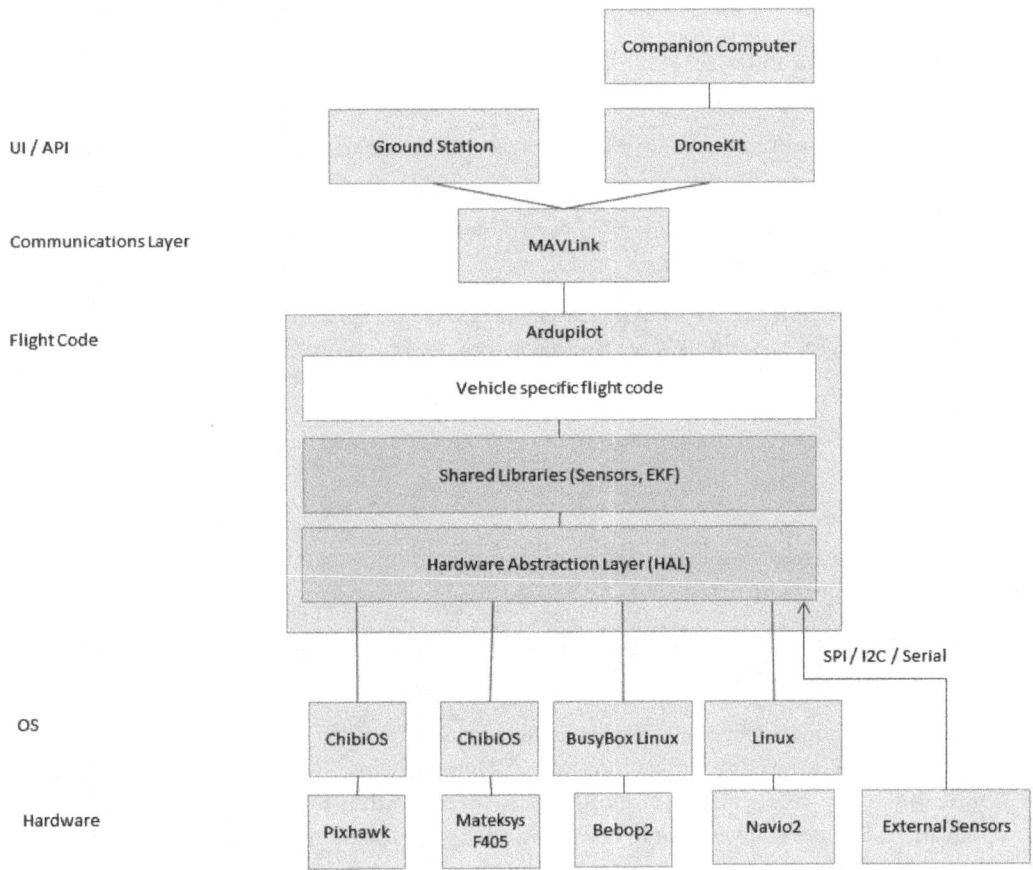

Figure 6.3 – High-level architecture diagram of ArduPilot

The ArduPilot flight control architecture is characterized by a monolithic design that integrates various components into a single firmware code base. The architecture is designed to support a wide range of UAVs, including multirotors, fixed-wing aircraft, helicopters, ground rovers, and boats. Here are key components and their roles within the ArduPilot architecture:

- **ArduCopter**: This module is responsible for the control and stabilization of multirotor vehicles, supporting different configurations such as quadcopters, hexacopters, and octocopters.
- **ArduPlane**: Designed for fixed-wing aircraft, ArduPlane handles flight control, navigation, and mission planning for various types of planes.

- **ArduRover**: Geared toward ground-based vehicles, ArduRover provides control and navigation capabilities for land-based drones, including rovers and boats.
- **Common code base**: ArduPilot maintains a common code base that supports various vehicle types. This integrated approach simplifies development and firmware updates.
- **Hardware Abstraction Layer (HAL)**: The HAL in ArduPilot abstracts the underlying hardware, allowing the firmware to be easily ported to different autopilot boards.
- **The AP_Motors module**: This module manages motor outputs and control signals, ensuring proper actuation for different vehicle types.
- **The AP_Notify module**: Responsible for system notifications, `AP_Notify` helps in providing status updates and warnings to users or GCSs.
- **The AP_Mission module**: Mission planning and execution are handled by the `AP_Mission` module, enabling autonomous waypoint navigation and mission scripting.

Though we have seen the structure of ArduPilot quite briefly, let's walk through the preceding architectural diagram for a brief understanding of the topics at various levels.

UI/API

UI/API refers to the external source via which the ArduPilot flight stack receives commands. Some examples of UI/API sources are as follows:

- **Ground station**: The ground station referred to here is GCS software, which is installed on a laptop or tablet and serves as a UI to configure and command the drone.
- **DroneKit**: DroneKit is a Python API that helps developers develop applications to communicate with the drone via the communications layer. DroneKit helps developers develop lightweight applications to pass certain commands to the drone rather than a complete GCS. These applications are useful if we want to communicate to the firmware via a developer board such as Raspberry Pi.

Communications layer

The communications layer is a part of the flight stack that is helpful in establishing communication links between the UI/API and the flight controller. It is through this communications layer that the GCS communicates with the drone flight code.

An example of a communications layer is as follows:

- **MAVLink**: As we covered earlier, MAVLink is a lightweight communication protocol that helps to transmit and receive commands between a GCS and a drone.

Flight code

The ArduPilot flight code refers to the firmware that runs on the autopilot hardware and controls the operation of UAVs. This code is written in C++ and encompasses a wide range of functionalities for different types of vehicles, including multirotors (ArduCopter), fixed-wing aircraft (ArduPlane), ground rovers (ArduRover), and boats. Here are more details of the flight code in ArduPilot:

- **Vehicle-specific code**: Vehicle-specific code in ArduPilot refers to the portions of the firmware that are tailored to the unique characteristics and requirements of different types of unmanned vehicles. ArduPilot supports various vehicle types, including multirotors (ArduCopter), fixed-wing aircraft (ArduPlane), ground rovers (ArduRover), and boats. Each vehicle type has its own set of dynamics, control mechanisms, and operational considerations, necessitating specific code to ensure optimal performance.

    Here's an overview of how vehicle-specific code is structured within ArduPilot:

    - **ArduCopter (multicopters)**: The ArduCopter code includes algorithms and control logic specific to multirotor vehicles. This encompasses stabilization, altitude control, GPS navigation, and features such as automated takeoff and landing. Vehicle-specific parameters, such as motor mixing and control gains, are defined in this section.
    - **ArduPlane (fixed-wing aircraft)**: The ArduPlane code focuses on the unique characteristics of fixed-wing aircraft. It incorporates algorithms for maintaining stable flight, waypoint navigation, and automated landing. Parameters related to airspeed control, navigation modes, and servo outputs are tailored for fixed-wing vehicles.
    - **ArduRover (ground rovers)**: The ArduRover code is designed for ground-based vehicles, such as rovers and cars. It includes control logic for steering, throttle, and braking. Navigation features, such as waypoint following and obstacle avoidance, are specific to ground-based dynamics.
    - **Boats and surface vehicles**: For boats and surface vehicles, the code addresses the challenges of water-based navigation. This includes controls for propulsion, steering, and stability on water surfaces.

- **Shared libraries**: ArduPilot is primarily written in C++ and employs a modular architecture to enhance code reusability and maintainability. However, the concept of shared libraries, as commonly understood in traditional software development, might not be directly applicable to ArduPilot. Instead, ArduPilot uses a modular approach with libraries and modules.

    In ArduPilot, shared libraries or modules refer to reusable code components that encapsulate specific functionalities. These modules are designed to be independent and interchangeable, allowing developers to build upon existing code and add new features without affecting the entire system.

- **Hardware abstraction layer**: The HAL in ArduPilot, known as `AP_HAL`, is a crucial component that abstracts and encapsulates hardware-specific details, allowing the autopilot software to run seamlessly across different flight controller hardware. `AP_HAL` provides a standardized interface for accessing peripherals, sensors, communication ports, and other hardware components. By employing a HAL, ArduPilot achieves portability, enabling the same code base to be compatible with various autopilot boards. This abstraction layer simplifies the development process, as developers can write code against a consistent API without worrying about the intricacies of the underlying hardware, facilitating a modular and scalable approach to UAV system design.

Operating system

Chibi Operating System (ChibiOS) was one of the **real-time operating systems (RTOSs)** used in the ArduPilot project. ChibiOS plays a crucial role as the underlying OS for certain autopilot boards, providing a framework for managing tasks, scheduling, and handling low-level hardware interactions. Here's a brief explanation of the role of ChibiOS in ArduPilot:

- **RTOS functionality**: ChibiOS is an open source RTOS designed for embedded systems, and it provides real-time capabilities essential for UAV applications. It facilitates task scheduling, interrupt handling, and efficient resource management, ensuring timely and deterministic execution of critical flight control processes.

- **Hardware abstraction**: ChibiOS serves as a HAL in ArduPilot, allowing the flight control code to interact with various sensors, peripherals, and communication interfaces in a consistent manner. This abstraction simplifies the development process and enhances portability across different autopilot hardware.

- **Concurrency and multithreading**: ChibiOS enables the execution of multiple tasks concurrently, allowing the autopilot to handle diverse operations simultaneously. This is essential for managing tasks such as sensor readings, control calculations, communication, and navigation concurrently.

- **Low-level hardware interaction**: ChibiOS provides low-level drivers and abstractions for interacting with specific hardware components, including timers, serial ports, and I2C buses. This allows ArduPilot developers to write code at a higher level without directly dealing with the intricacies of hardware registers and configurations.

- **Portability**: One of the key advantages of using ChibiOS is its portability across different microcontroller architectures. ArduPilot can leverage ChibiOS on a variety of autopilot boards, ensuring flexibility and compatibility with diverse hardware platforms.

External sensors

In ArduPilot, external sensors refer to additional sensors that can be connected to the autopilot system to enhance its perception and data collection capabilities. These sensors can provide supplementary information beyond what is integrated into the main flight controller, enabling more accurate navigation, control, and decision-making. Examples of external sensors in ArduPilot include the following:

- **LiDAR or laser range finders**: LiDAR sensors are commonly used to measure distances by emitting laser beams and measuring the time it takes for the beams to return. They provide high-precision altitude information and assist in terrain following and obstacle avoidance. An example is the LightWare SF20 LiDAR.

- **Radar altimeters**: Radar altimeters measure the distance from the vehicle to the ground and are particularly useful for accurate altitude control, especially during low-altitude flights. They contribute to safer landings and takeoffs. An example is the TeraRanger Tower.

- **Optical flow sensors**: Optical flow sensors use visual information to estimate changes in the vehicle's position. They are beneficial in scenarios where GPS signals may be unreliable, such as in indoor environments or during low-altitude flights. An example is the PX4Flow optical flow sensor.

- **GPS/compass modules**: While GPS is integrated into many flight controllers, external GPS modules with integrated compasses provide redundancy and improved accuracy in orientation estimation. Examples include the u-blox M8N GPS module.

- **Air data sensors**: These sensors measure airspeed, barometric pressure, and temperature, providing crucial data for accurate altitude control and airspeed estimation. An example is the MS5611 barometer.

- **Magnetometer modules**: External magnetometers can be used to obtain more accurate compass readings, reducing the impact of magnetic interference from other onboard electronics. An example is the HMC5883L magnetometer.

Summary

In this chapter, we have gone through different firmware and covered the structure and architecture of PX4 and ArduPilot, firmware that would help you to potentially know about the building blocks of the flight stack of the drone. We have also referenced the open source firmware stack PX4 and ArduPilot, which are leaders in the drone world.

In the next chapter, we will learn about the GCS and its operation, which will help in configuring the drone and flying autonomous missions.

7
Introduction to Ground Control Station Software

In previous chapters of this book, we studied major unmanned systems, their subsystems, and the sensors on the hardware side. We also studied firmware, which is used as a link between hardware and software to make the system function, as well as the basics of command and control software, called ground control software, with its different types and usage. In this chapter, we will learn about the in-depth usage of ground control software that is used to command, control, and set up a drone for flight. A **Ground Control Station** (**GCS**) is also used to transmit and receive real-time telemetry data during the flight and monitor the drone's navigation and sensor health while it flies in both manual and autonomous modes. We will walk through the different kinds of GCS systems and how we can control and configure the drone with the help of a GCS.

In this chapter, we are going to cover the following topics, which will be helpful for us to understand GCSs and their usage in drones. By learning this, you will be able to download and install a GCS on your desktop or mobile, connect it to the drone that you have built, configure it, and fly in the mode you want.

The following are the main topics to be covered in this chapter:

- Introduction to GCSs
- Major GCS software that is available on the market
- AeroGCS software overview
- Methods of connecting a drone
- Autonomous mission planning

Technical requirements

To install and run the GCS software discussed, it's advised to have a computer/desktop with an x64-based i5 1.9 GHz CPU or better with a minimum of 30 GB free space and 8 GB RAM. It's recommended to have a 1 GB graphics card and a USB 2.0 slot. We will be using a laptop/desktop with the same configuration to download and install the GCS software.

The computer/laptop requires a minimum of Windows 10 or Ubuntu 20.xx LST versions. The system should have LAN or Wi-Fi connectivity and 1,366 x 768 landscape resolution.

If you're planning on using a mobile-based Android version, then it should have a USB tethering feature or Wi-Fi.

Introduction to GCSs

GCS software plays a pivotal role in the operation of UAVs by providing a centralized interface for users to interact with and control their drones. By serving as a bridge between operators and the airborne vehicle, GCS software enables real-time monitoring, mission planning, and data analysis. These applications are essential tools for both hobbyists and professionals, offering a user-friendly platform to command and oversee the UAV's flight. GCS software typically provides features such as live telemetry, waypoint navigation, camera control, and the ability to visualize and analyze flight data. Whether for recreational drone enthusiasts, researchers, or commercial drone operators, well-designed GCS software enhances the user experience, ensuring safe, efficient, and controlled UAV missions.

There are various kinds of GCS software available on the market, catering to different user needs, from hobbyists to professional operators. Here are some common types:

- **Open source GCS**: Open source GCS software is freely available and often developed collaboratively by the drone community. It offers features such as mission planning, telemetry visualization, and parameter tuning. Mission Planner is commonly used with ArduPilot-based drones, while QGroundControl is popular for PX4-based vehicles.

    Examples: Mission Planner and QGroundControl.

- **Proprietary GCS**: Proprietary GCS solutions are developed by drone manufacturers for their specific hardware. DJI's GCS software, such as DJI GO and GS Pro, is tailored for their drone platforms. These applications often provide seamless integration with the manufacturer's hardware and offer advanced features such as intelligent flight modes and professional cinematography tools.

    Examples: DJI GO and DJI GS Pro.

- **Cloud-based GCS**: Cloud-based GCS solutions allow users to plan, monitor, and analyze drone missions through online platforms. These services often offer features such as flight logging, mission planning, and data analytics. DroneDeploy, for instance, enables users to plan flights, capture aerial imagery, and process data for mapping and surveying applications.

    Examples: Airdata UAV and DroneDeploy.

- **Mobile GCS apps**: Mobile GCS apps run on smartphones or tablets, providing a portable solution for drone operators. Litchi offers advanced mission-planning tools for DJI drones, while PIX4Dcapture is designed for mapping and photogrammetry missions. These apps typically feature intuitive interfaces and real-time telemetry.

  Examples: Litchi and PIX4Dcapture.

- **Enterprise GCS**: Enterprise GCS solutions cater to professional and industrial drone applications. They often include advanced mission-planning tools, fleet management, and integration with other enterprise systems. UgCS, for example, supports a wide range of UAV platforms and offers features such as geofencing and multiple vehicle coordination.

  Examples: UgCS and PrecisionFlight.

- **Simulation GCS**: Simulation GCS tools allow users to practice and test drone missions in a virtual environment. These applications are valuable for training purposes and refining mission plans before actual flight operations.

  Example: Mission Simulator.

Hence, we have seen the different types of GCSs. In the following sections, we will study the various GCSs available on the market.

Major GCS software that is available on the market

In the rapidly evolving landscape of UAVs, the availability of sophisticated GCS software has become instrumental in shaping the success of drone operations. Numerous GCS solutions cater to a diverse user base, ranging from hobbyists and enthusiasts to professional operators and enterprises. This overview explores some of the major GCS options available on the market, each offering distinct features and capabilities to enhance the planning, control, and monitoring of drone missions.

Mission Planner or APM Planner

Mission Planner or APM Planner is a piece of open source software supported by the open source drone community and developed by Michael Oborne and is especially meant for drones powered by ArduPilot firmware (`www.ardupilot.org`). This GCS is very powerful and can be configured with a plane, copter, and rover. Currently, this software is compatible with Windows only but can be configured to Linux machines by installing the module called **mono**.

Mission Planner can be downloaded from `https://firmware.ardupilot.org/Tools/MissionPlanner/MissionPlanner-latest.msi`.

Introduction to Ground Control Station Software

The following screenshot shows the standard dashboard (home screen) of Mission Planner:

Figure 7.1 – Dashboard (home screen) of Mission Planner GCS

Major features

Mission Planner is a versatile piece of GCS software that gives you complete control over your system, right from configuring and tuning to flying and debugging. It offers you an added range of additional peripherals that you can attach to your drones as per the application requirements so you can configure and test them. It primarily focuses on providing tools for mission planning, vehicle monitoring, and configuration management. Here are some common features of Mission Planner:

- **Flight planning**: It gives the advantage of planning autonomous flight missions by specifying waypoints, altitude, and other parameters. Complex missions can be designed based on this.

- **Real-time telemetry**: Real-time telemetry data can be seen on the screen, which includes the vehicle's position, altitude, speed, and sensor health information. This helps users monitor the status of their vehicle during operation.

- **Configuration and calibration**: Mission Planner is used to configure and calibrate the various parameters of vehicles, such as radio settings, motor and sensor calibration, and other hardware configurations.

- **Data logging**: Mission Planner enables the logging of flight data, which allows users to review and analyze the performance of their vehicles after a mission. This data can be crucial for troubleshooting and improving the vehicle's capabilities.

- **Firmware updates**: Users can update the firmware of their vehicles through Mission Planner. Firmware updates often include improvements, bug fixes, and new features.

- **Camera control**: Mission Planner allows users to control and configure the camera settings of vehicles equipped with cameras or gimbals. This is particularly useful for aerial photography and mapping missions.
- **Geofencing**: Mission Planner supports the creation of geofences, which are virtual boundaries that restrict the movement of the vehicle. This is useful for ensuring that the vehicle stays within a specified area.
- **Simulation**: Users can simulate missions in a virtual environment before executing them in the real world. This helps in testing and refining mission plans without the risk of damaging the vehicle. The simulation gives a 2D view of the vehicle and environment with live data from sensors.
- **Joystick support**: Mission Planner is often compatible with joysticks and other input devices, allowing manual control of the vehicle when necessary.
- **Mission analysis**: After a mission, users can analyze the recorded data, including the flight path, altitude, and other relevant information. This helps in evaluating the success of the mission and identifying areas for improvement.

Apart from the key features mentioned previously, Mission Planner also gives you some advanced tools such as drone swarming and MAVLink packet signing for extra security. This GCS software provides the flight logs and analysis in one place without the requirement of any third-party software for the log analysis part.

To learn more about the configuration, setup, and flight planning in Mission Planner, you can check out the ArduPilot page.

QGroundControl

QGroundControl is another famous GCS managed and developed by the open source drone community for the full flight control and setup of PX4 and ArduPilot vehicles. Just like any other GCS software, QGroundControl provides an easy approach for setting up a vehicle, right from the installation of firmware to flight logging and analysis. QGC is compatible with Windows, Linux, and Android and can be downloaded from http://qgroundcontrol.com/.

The basic principle of QGroundControl is the same as other drone systems in which MAVLink is the communication protocol and the flight stacks are ArduPilot or PX4.

The major features of QGroundControl include the following:

- Supports any flight stack that supports the MAVLink protocol
- Provides the full setup and configuration of the system
- Aids in flying both manually and autonomously
- Supports multiple maps, such as Bing and Google Maps, for mission planning

- Support for built-in video transmission
- Compatible with almost all operating systems
- Can be used for various applications such as surveys, surveillance, and agriculture

The following screenshot shows the dashboard or home screen of a QGroundControl station:

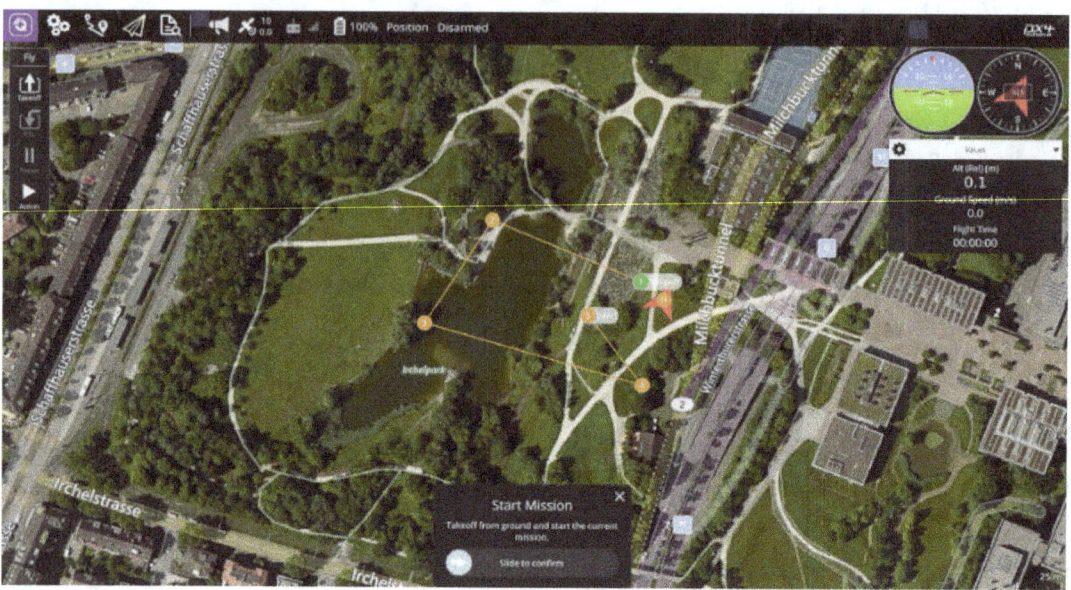

Figure 7.2 – QGroundControl home screen in mission-planning mode

This shows the flying window and a planned mission that is ready to start.

UgCS

UgCS is an advanced piece of GCS software that is used for different types of drone mission planning and execution. UgCS is not open source software; rather, it is proprietary software that requires a license to use with an annual fee; however, there is a free version available with limited features for the user. This software is compatible with multiple flight stacks, such as ArduPilot, PX4, DJI, Parrot, and Autel; hence, it supports multiple types of drones. UgCS is compatible with Windows and macOS.

UgCS can be downloaded from `https://www.sphengineering.com/support`.

Major GCS software that is available on the market   125

The following screenshot shows the UgCS mission-planning software, indicating drone system health, telemetry data, and a mission plan:

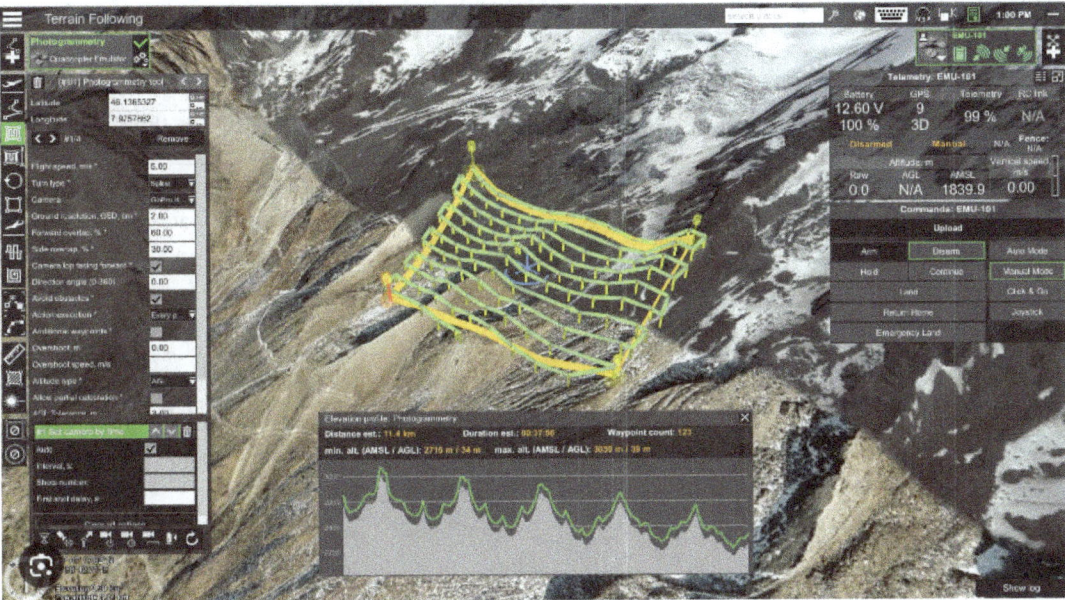

Figure 7.3 – A screenshot of the UgCS mission-planning software

UgCS has unique features, has a user interface, and is simple to operate, which sets this GCS apart from other control stations. A few of its major features are listed as follows:

- Can create linear or circular waypoints with area and perimeter scanning
- Supports 2D and 3D maps for easy mission views and terrain following
- Supports particular LiDAR tools for LiDAR sensors
- Enables multiple SDK connections for custom development
- Great 3D mission-planning environment

The following is a brief comparison of Mission Planner, QGroundControl, and UgCS in tabular format:

Feature	Mission Planner	QGroundControl	UgCS
Platform Compatibility	Windows (primarily), adaptable to other platforms	Cross-platform (Windows, macOS, Linux, and Android)	Windows, macOS, and various Linux distributions
Development Status	Community-driven, actively maintained on GitHub	Open source, actively developed by the PX4 community	Commercial software with updates and support
Integration	Associated with ArduPilot firmware	Developed with the PX4 flight stack, supports others	Compatible with various autopilots, including DJI
Autopilot Support	Primarily ArduPilot-based vehicles	Designed for PX4-based vehicles, supports others	Compatible with a wide range of autopilots
User Interface	Comprehensive interface with various tools	User-friendly, emphasis on simplicity	Intuitive interface balancing simplicity and depth
Community Support	Active community contributions on GitHub	Supported by the PX4 community	Commercial software with user support
Advanced Features	Emphasis on mission planning, telemetry, config	Balances simplicity with advanced features	Advanced features such as photogrammetry and agriculture
Hardware Support	Broad support, especially for ArduPilot-based UAVs	Compatible with a variety of autopilots	Supports various autopilots, including DJI and ArduPilot
License	Open source (GPLv3)	Open source (Apache 2.0)	Commercial with a free version available
Use Case Focus	Mission planning, telemetry, configuration	Mission planning, telemetry, simulation	Professional applications, advanced features

Table 7.1 – Comparison table of Mission Planner, QGroundControl, and UgCS

Apart from the preceding GCS software, there are many more that are available for commanding and controlling drones, but we have selected these pieces of software for a basic overview without going into too much depth. However, the basic functionality of these pieces of software is similar. In a later section, we will look at another proprietary GCS, named AeroGCS, in depth, understanding how we can set up and configure a drone using this software. We will study AeroGCS as a GCS and learn how we can configure the drone and plan successful missions with it.

AeroGCS software overview

AeroGCS is a proprietary GCS developed by the AeroGCS company, based in India, for drone configuration, piloting, fleet management, and drone project management. This software is available in multiple versions and can exist on a laptop or your mobile phone. The great part about AeroGCS software is that it greatly helps pilots plan missions and works in a cloud-based environment, giving access to the data anytime and anywhere. By using this, you can store your mission data on the cloud and monitor different drones working together on different projects.

Like other GCS software, AeroGCS helps to plan missions, set parameters, and have complete command and control over a drone. AeroGCS software is based on the MAVLink protocol; hence, it is compatible with PX4, ArduPilot, or any MAVLink-based flight stack. However, the PX4 and ArduPilot flight stacks are most preferred due to their simplicity and rich user interfaces.

To download the software, visit `https://aeromegh.com/aerogcs-kea/`.

AeroGCS main dashboard

The dashboard is the primary AeroGCS window, containing the settings of different parameters with specified values and conditions that are required to operate remote pilot aircraft. *Figure 7.4* shows what we see after logging in to the software. Now, we will go through the different shortcuts for handling drone operations and the various information located on the dashboard.

Let's look at and explore each item on the dashboard to understand its functions and features:

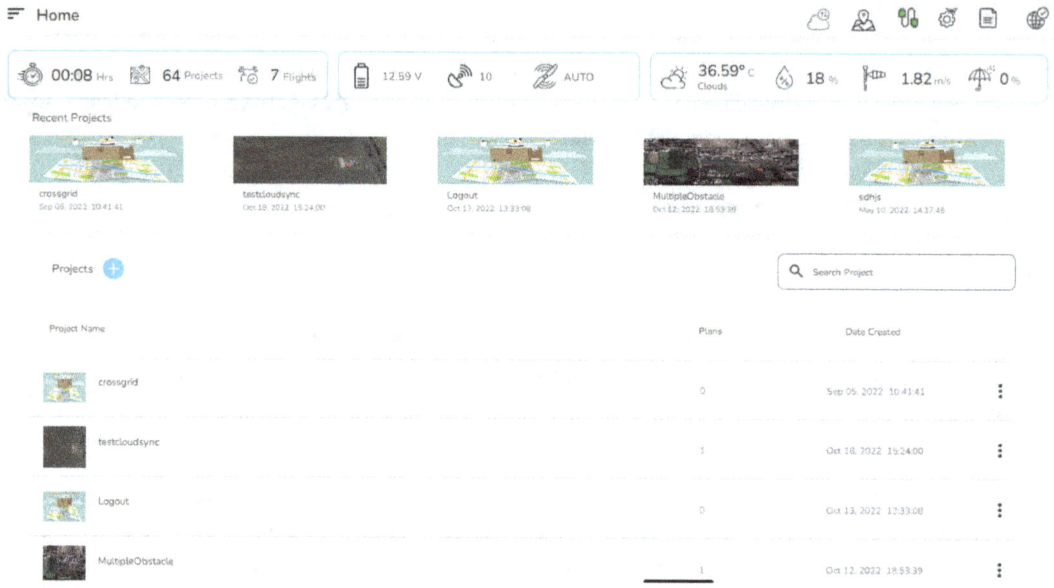

Figure 7.4 – AeroGCS dashboard

The top bar in the upper-right corner looks like this:

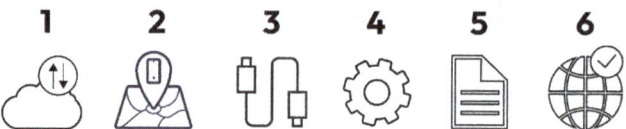

Figure 7.5 – Top bar options in the AeroGCS dashboard

The items are as follows:

1. **Cloud sync**: This feature is useful for on-field pilots to manage and upload mission plans, captured images, and so on to the cloud directly so that the information can be shared with team members directly through the **Team Access Management** features included in the **AeroMegh** cloud service platform. This helps ensure the manager and team members are looking at up-to-date information for ongoing projects.

2. **Direct Fly View**: The **Direct Fly View** button takes you directly to the navigation bar, where the pilot can see data related to the drone, such as latitude, longitude, speed, and orientation.

3. **Connect a device**: This button helps to connect the GCS and the drone. This is a crucial step; the drone should be completely connected to the software before a flight. We need to select the best protocol that we can use (for example, serial, TCP, or UDP) for drone connection. Different baud rates can be selected as per the application.
4. **RPA configuration**: Further drone settings can be accessed with this button. RPA here refers to remotely piloted aircraft.
5. **Notifications**: The notifications button helps us to see real-time messages regarding everything that is happening during the flight, such as warnings, flight logs, and sending and receiving commands. This notification also includes critical messages such as sending the flight plans, arming motors, and taking off and landing.
6. **Internet connectivity**: The internet connectivity option shows us whether AeroGCS is connected to the internet or not; if not, then the icon is struck through.

AeroGCS home menu

The AeroGCS home window is the key navigation window that is used to configure the different functionalities of the drone. The following is the home menu, which shows the different options available to the user:

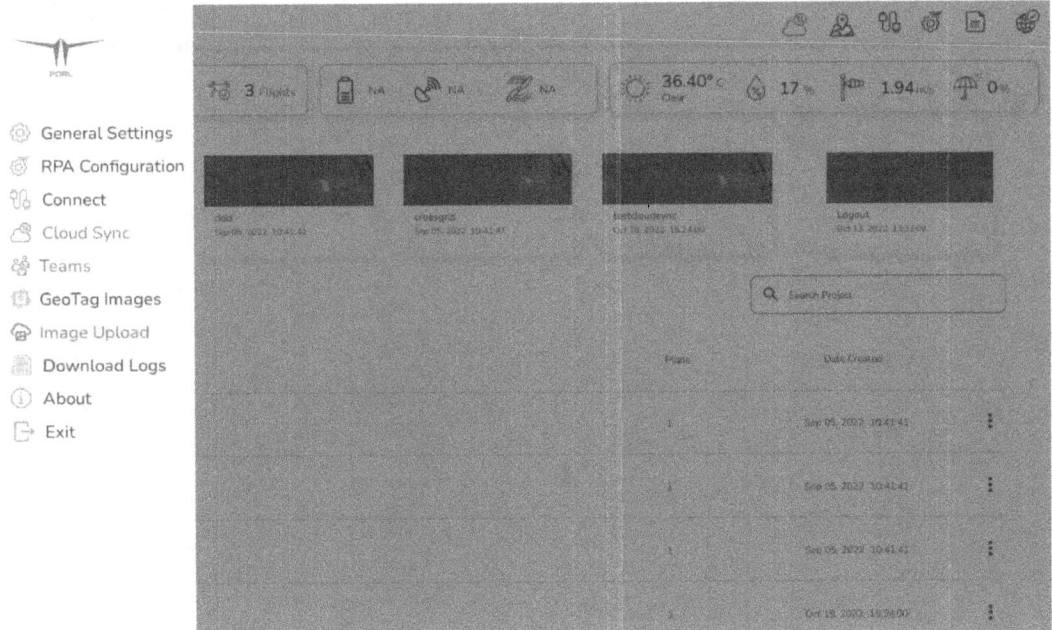

Figure 7.6 – The home menu of AeroGCS

We will take a brief look at the **General Settings** option here.

The general settings refer to the basic operational settings of the AeroGCS software; you can use this to create your personal profile, select the theme of your choice, see weather settings, and change the measurement units as required. The following are the options available in **General Settings**:

- **AeroMegh login**: Since AeroGCS is proprietary software, an annual license is required to log in to access a user's profile and manage this profile and projects. The following is the login page:

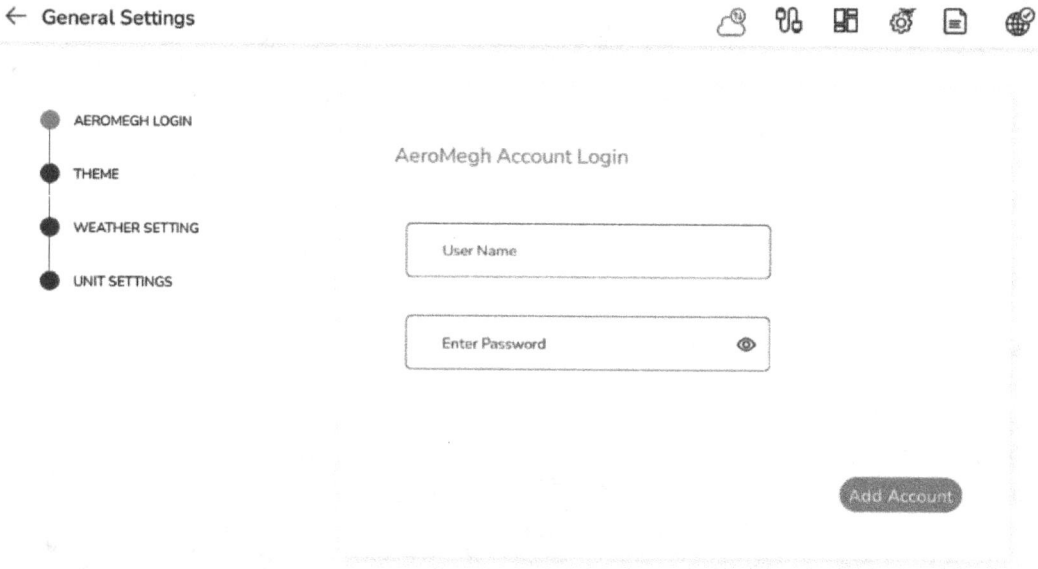

Figure 7.7 – Login window of AeroGCS

- **Theme**: Users can change the theme from this setting. The default theme is **AeroGCS Light**, which can be changed to **AeroGCS Dark**. The application must be restarted for the changes to take place. The following is a screenshot of the theme change option:

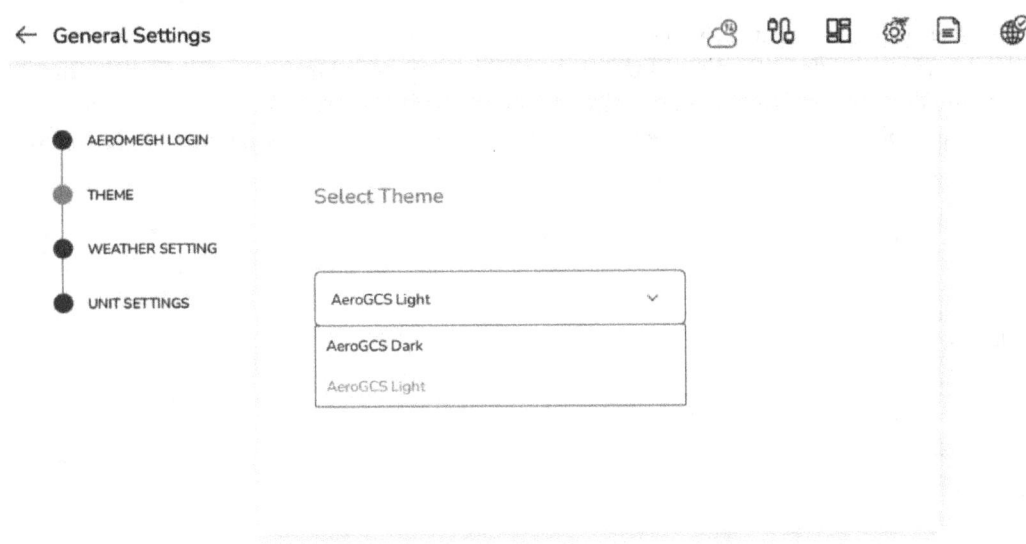

Figure 7.8 – Theme selection in AeroGCS

- **Weather setting**: The weather settings can be applied by accurately selecting the city in the tab. This considers the current weather:

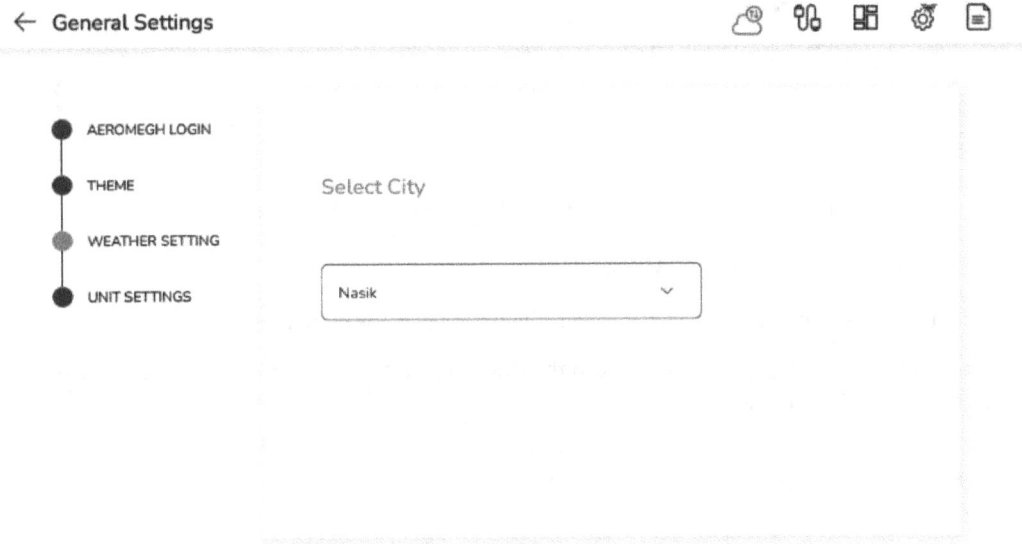

Figure 7.9 – Options to select a city for weather forecasts

- **Unit settings**: Unit settings are required prior to the flight as per the convenience of the pilot and system to reflect the desired unit measurement system. Unit settings cover the major parameters required by the pilot, such as distance, area, speed, and temperature. Different units are followed for these conventions across the globe; hence, this can be useful for pilots across the globe:

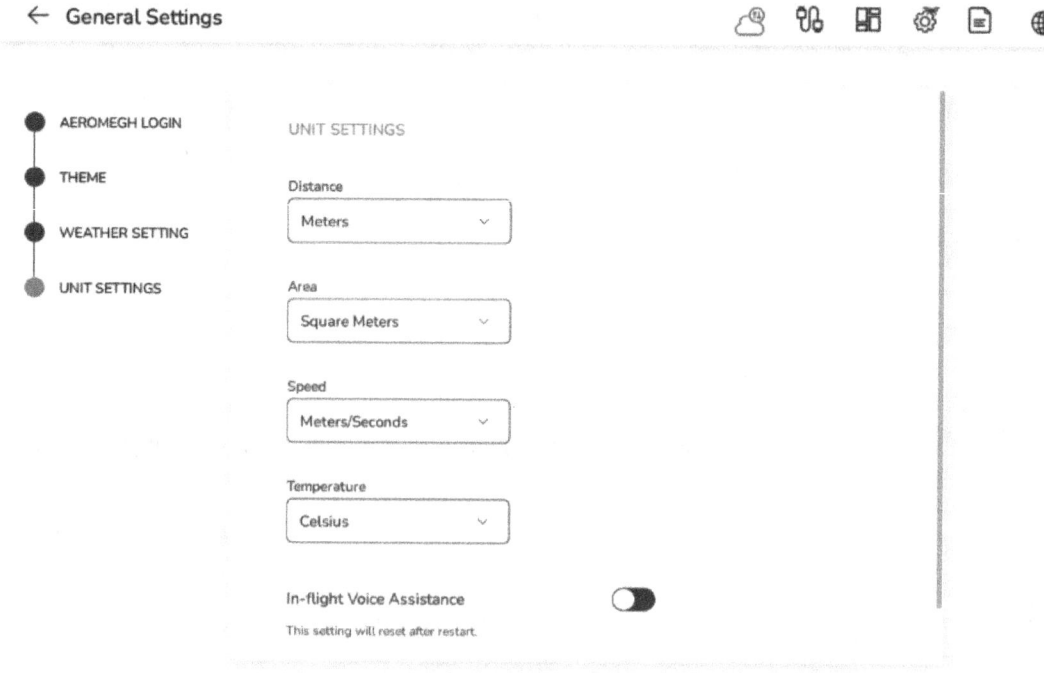

Figure 7.10 – Selection of units for different physical quantities

The user can set units for the following:

- **Distance**: The user can set the distance units to either **Meters** or **Feet**
- **Area**: The units of the area can be set to the desired unit from the drop-down list, as shown in the following screenshot:

AeroGCS software overview 133

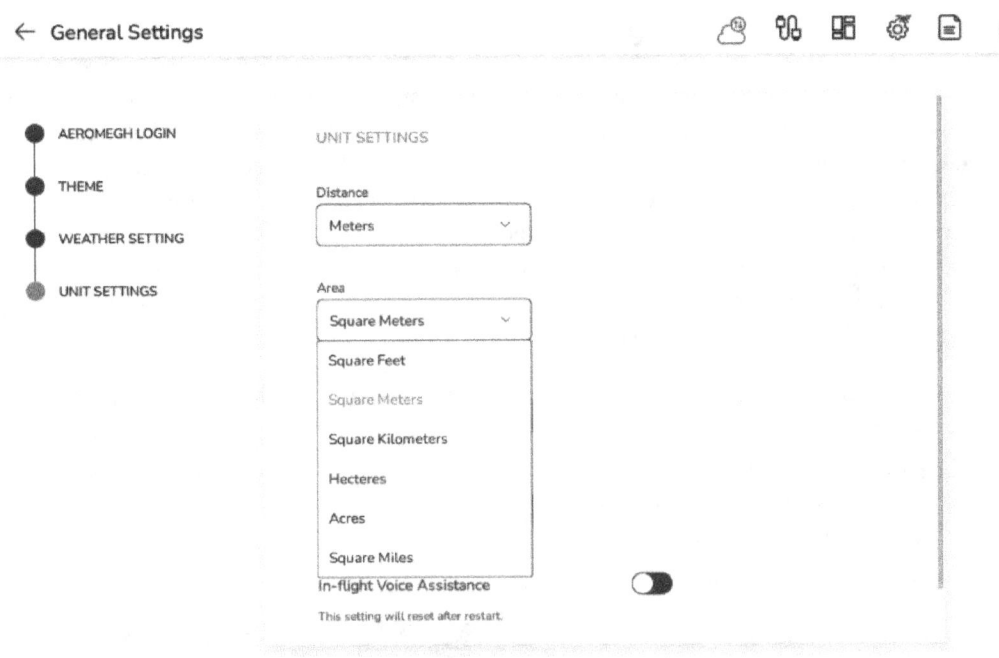

Figure 7.11 – Selection of units

- **Speed**: The vehicle's speed units are set according to the requirements
- **Temperature**: Units of temperature can be set to **Fahrenheit** or **Celsius**

Hence, from the preceding introduction, we have seen how to configure the general settings of AeroGCS. Now, we may proceed to understand how to configure **remotely piloted aircraft** (**RPA**) or drones. We will not cover the full setup of it via AeroGCS; we will only cover the basic and introductory parts. To understand this software completely, you can visit their website (www.aeromegh.com) and access their tutorials.

Methods of connecting a drone

Connecting a drone to the GCS is the primary step that should be followed before proceeding with any further steps. As a common understanding, we can only access the internal parameters and program the drone once we are connected to it via either a USB cable or a radio frequency module. There should be proper communication between the drone, the GCS, and the remote control. It is incredibly important for the pilot to provide the right settings and methods to connect the drone, which we can see in the following screenshots. Additionally, it is highly recommended that you read and go through the manufacturer manual or datasheet for the drones you are using. We will go ahead with an RFD900 model and a serial connection:

1. Click the **Connect** icon on the top bar of the dashboard (or from the home menu) to connect to the drone.
2. We need to connect to the drone before we start a project. Say yes to any prompt that asks you to connect. In the following screenshot, we can see the third option for connecting a drone to the GCS (laptop/tablet/computer). We already studied different ways to connect the drone in previous chapters, such as with RFD/USB:

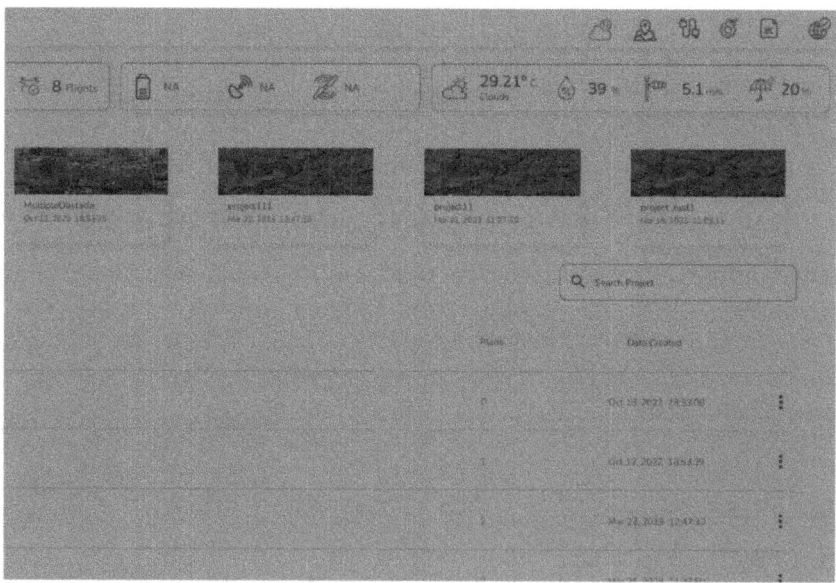

Figure 7.12 – Connect options

In the following screenshot, we can see a popup; the software requests to connect a device to it. Once you have clicked on **Yes**, the next steps to connect the drone to the GCS via a radio module will appear. This will ask you to select the protocols that we want to connect the drone to and various other parameter configuration options:

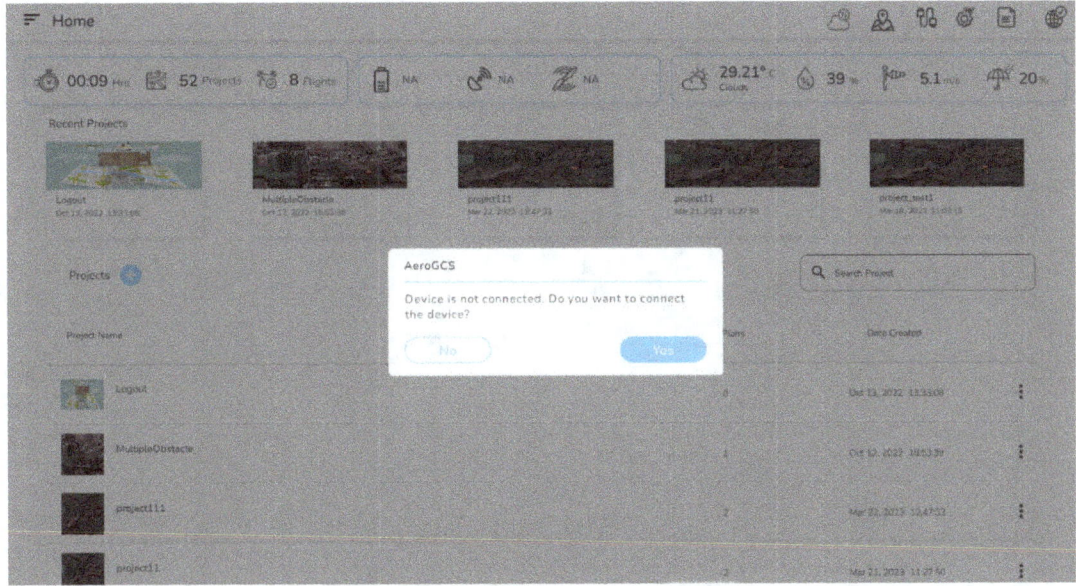

Figure 7.13 – Prompt to connect to a drone

The following are the steps for establishing the connection for any of these methods:

1. Choose a communication link. It could be **Serial**, **TCP**, or **UDP**. Select the suitable link based on the communication device you choose, as shown in the following screenshot. Since we have mentioned the use of RFD 900, we will go ahead with choosing the **Serial** connection:

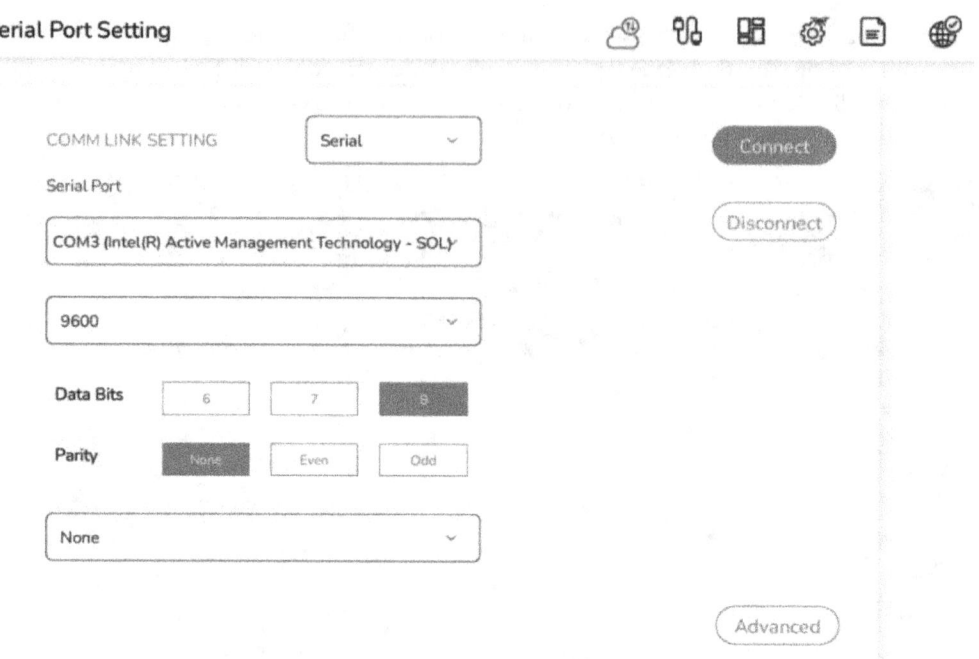

Figure 7.14 – Connection settings

Since we have chosen the serial connection, we will go ahead and look at the settings in relation to this option.

2. The **Serial Port** setting is required when we are connecting the drone using a serial protocol, such as UART, and serial radio frequency modules, such as RFD 900 433Mhz, via USB. Since these devices work on serial protocols, **Serial** has to be selected at the top:

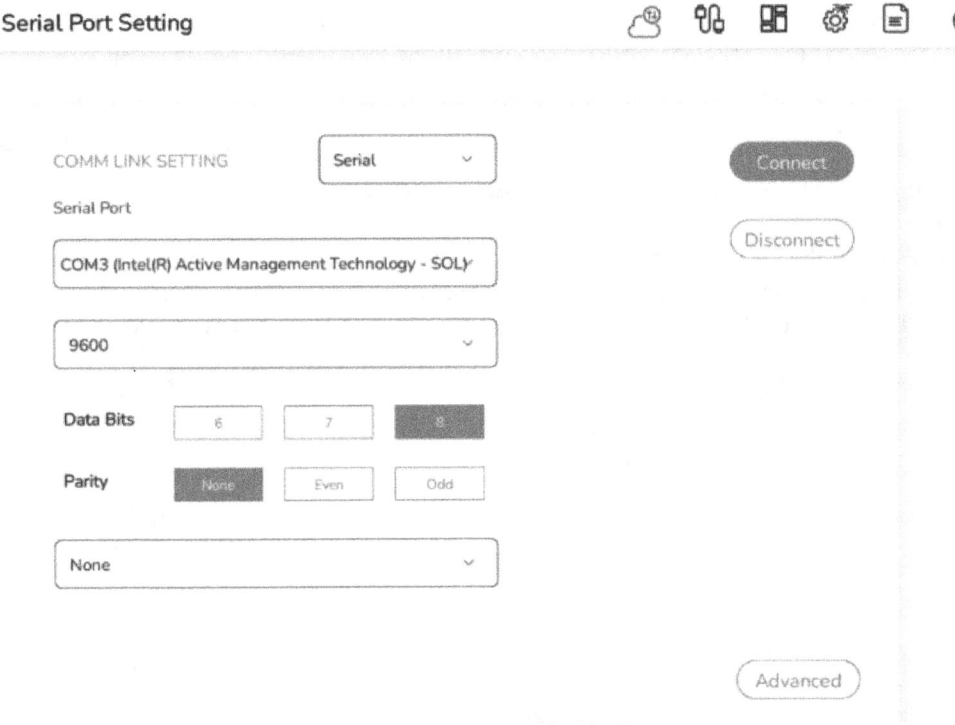

Figure 7.15 – Serial Port Setting

There are some parameters that need to be set before proceeding, which are as follows:

- **Serial Port**: When we connect the RFD 900 or any other module via USB, it establishes its own communication port through which the module can be accessed. This communication port will appear on the drop-down menu of **Serial Port**.

- **Baud rate**: We know that in serial communication, data is communicated via the transmission of bits (digital data) from one device to another through transmission media. The baud rate is the rate at which the number of signal elements or changes to the signal occurs per second when it passes through a transmission medium. The higher the baud rate, the faster the data is sent/received. For a wired communication via a USB cable, the baud rate is mostly 115,200, and for wireless connections, it is 57,600 if we are connecting to the flight controller.

- **Data bits**: Any messages that are transmitted over a serial port area sent via data bits. These can be a command, a sensor reading, an error message, and more. The majority of serial ports use 5 to 8 data bits for transmission, and text data is transmitted using 7 or 8 bits. These bits are used so that characters in the ASCII format can be used to represent the data.

- **Parity**: Parity is an error-checking procedure that uses the number of ones included in the transmitted bits. The number of ones must always be the same, either odd or even, for each group of bits that are transmitted without error.

Click on **Connect** on the top right of the window, and the drone should connect to the GCS.

3. **Connection via TCP and UDP**: If we are using a smart radio, such as Doodle Labs or Microhard, which we looked at in previous chapters, we can use a TCP/UDP connection to connect to a network. In the TCP connection, the user can set the host address, listing port, target address, and target port of the host module, and the drone will be connected via TCP. Hence, for **COMM LINK SETTING**, shown in the following screenshot, **TCP** now has to be selected:

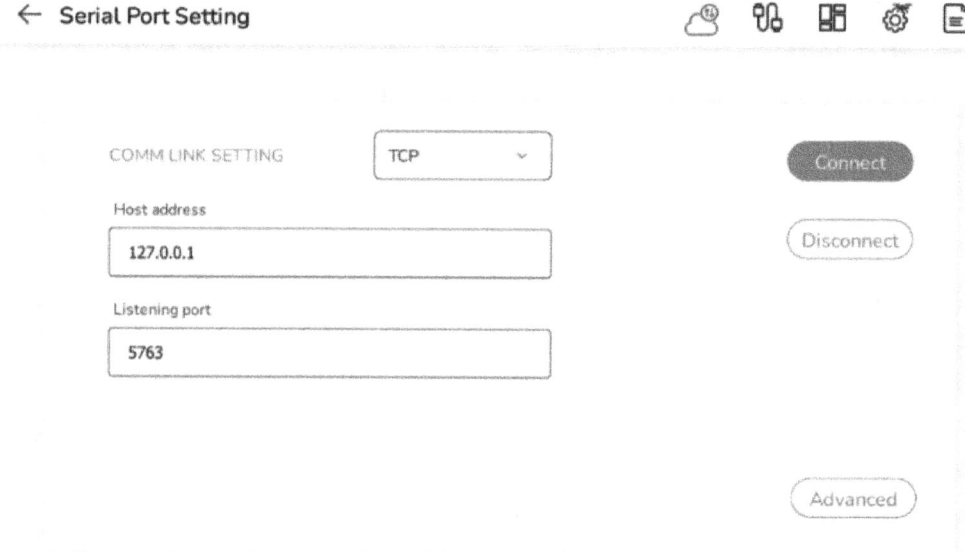

Figure 7.16 – TCP connection in AeroGCS

UDP stands for **User Datagram Protocol**, which is a transport layer protocol in the OSI model. It is not very reliable for data communication because in UDP, each packet is sent independently, without guarantees of delivery or orders; however, TCP provides a reliable connection to the system as it guarantees that data sent from one end is received correctly by the other end. In AeroGCS, a TCP connection is preferred over a UDP connection.

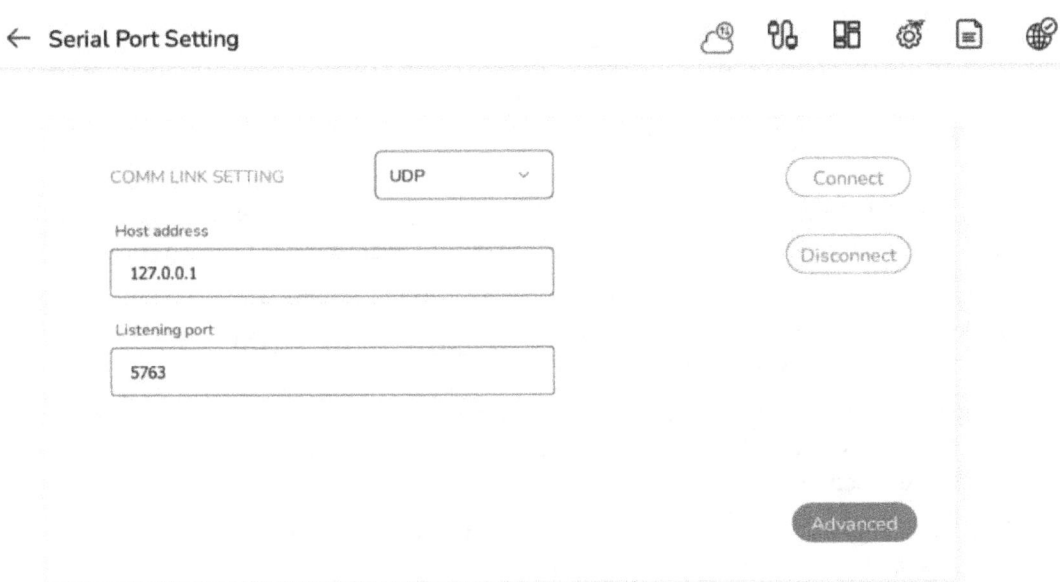

Figure 7.17 – UDP connection in AeroGCS

Hence, in the previous examples, we have seen the different methods of connecting to a drone (flight controller) through different protocols, which have to be selected accordingly. This will help us to establish a connection between the GCS and the drone. We can now command and configure the drone.

RPA or drone configuration

When we connect a drone to the GCS, it's mandatory to set up the necessary configurations of the drone and do the required calibration before any flights. The configuration of the drone is required so that the flight controller knows the exact values of the sensors, the airframe type, the battery status, the communication strength, and so on, which are required during the flight. The flight controller makes necessary decisions based on this data.

Checking sensor calibration, the motors, the ESCs, and the compass before a flight is important; this should be done before any flight.

The camera parameters should be set to capture the maximum number of best-quality images and videos of the area over which the drone is flying.

Necessary failsafe actions are to be taken for the emergency recovery of the drone in the case of a battery failure or a loss of communication.

The parameters related to the battery, power, and others are adjusted and set very easily with the RPA configuration.

Some important parameters to be set are as follows:

- **Selection of airframe**: Airframe selection is necessary, as it lets the flight controller know the type of drone (we have built a quadcopter, hexacopter, and octacopter, which are defined under the frame class).

    Next, the configuration of quad/hexa/octa and other similar drones needs to specify whether an X shape, + shape, or V shape is being used. These options are given in the drop-down menu.

    It's mandatory to reboot the flight controller once we have selected the airframe for the changes to take effect. Due to this selection, the ESC gets activated, and the basic physics of the drone is calculated by the flight controller for a stable flight.

- **Safety parameters**: Safety parameter settings are important to recover the drone in case of an emergency. These are some of the failsafe measures that are preloaded into the drone in case of a failsafe situation.

- **Failsafe**: Failsafe means if any of the crucial components of the drone stop working or malfunction, then what action must be taken by the drone itself for the safe recovery of the drone? These failsafe actions can be selected in this menu before the flight.

We will study the basic and important failsafes here that need to be set up before the flight:

- **Battery failsafe**: It triggers when the battery percentage of the drone falls to an already-set percentage, for example, 20%. We can choose an action for what the drone needs to do in that case.

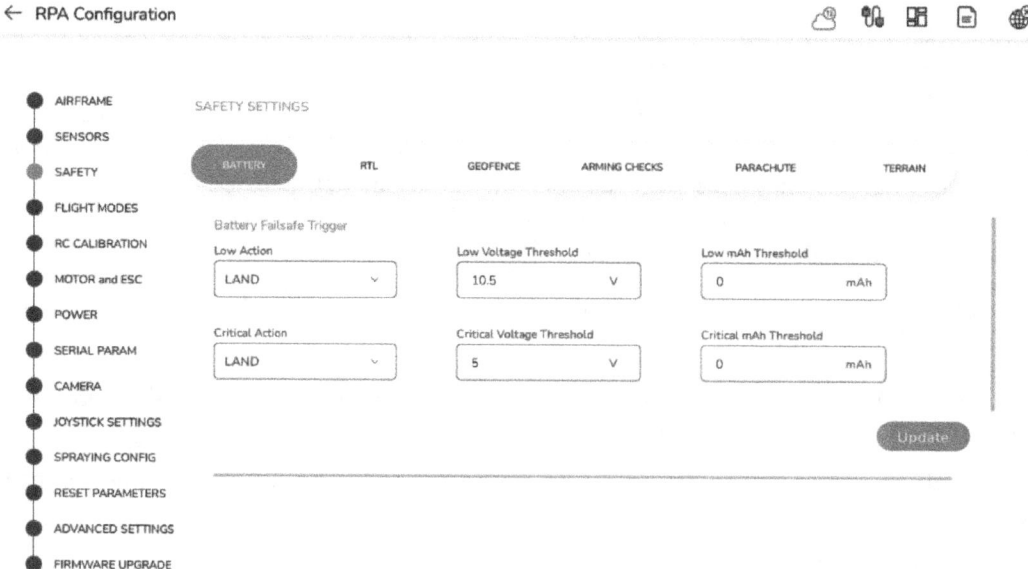

Figure 7.18 – Different battery failsafes

The following are some of the battery failsafes and their actions:

- **Low Action**: **None, LAND, RTL, SmartRTL, SmartRTL or Land, Terminate**
- **Critical Action**: **None, LAND, RTL, SmartRTL, SmartRTL or Land, Terminate**
- **Low Voltage Threshold**: Battery voltage that triggers the low action
- **Critical Voltage Threshold**: Battery voltage that triggers the critical action
- **Low mAh Threshold**: Battery capacity that triggers **Low Action**
- **Critical mAh Threshold**: Battery capacity that triggers **Critical Action**

We can even choose to activate or disable the failsafe action based on the needs of the applications.

- **Ground control failsafe triggers**: This failsafe triggers when the connection between the GCS software and drone breaks for any reason. The reason may be the hanging of the software, loose connections, or a loss of power. You can choose the action that is required when this happens, such as land or return to launch.
- **Throttle failsafe**: This failsafe triggers when the connection between the remote control and drone breaks due to any reason. The reason may be the hanging of the loose connections or a drone moving too far away from the range of the remote controller:

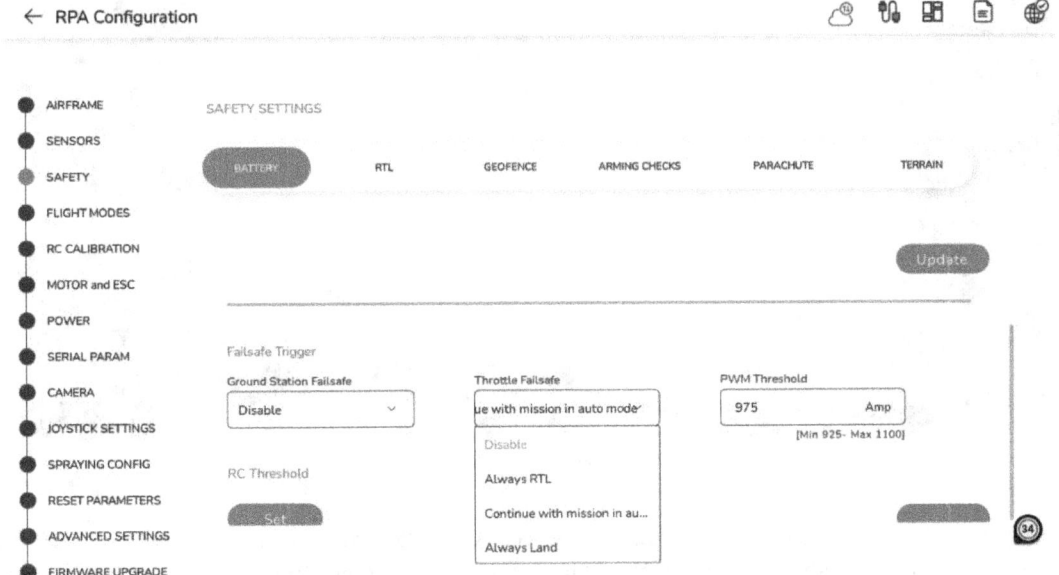

Figure 7.19 – Configuration for throttle failsafe

You can choose the action that is required when this happens, such as land or return to launch.

These are the main failsafes that can be configured. These failsafes must be set prior to the flight for the safety of the drone. There are some more advanced failsafes, such as geofence and GPS failsafes.

Autonomous mission planning

Mission planning is a key feature that every GCS offers. This feature helps a drone pilot program the drone with the flight paths they decide based on the environment and applications to make the drone follow it in completely autonomous mode without any pilot intervention. Mission planning requires the support of maps and a GPS to accurately plot the waypoints through which the drone will move and, hence, form a flight path.

A flight path is a set of longitudes, latitudes, and elevations (waypoints) used by a drone to automatically navigate. AeroGCS offers 10 preloaded flight paths, which can be chosen by the pilot as per their needs; hence, this makes the job of the pilot easy and accurate. In the following screenshot, we can see the different kinds of flight paths that AeroGCS offers:

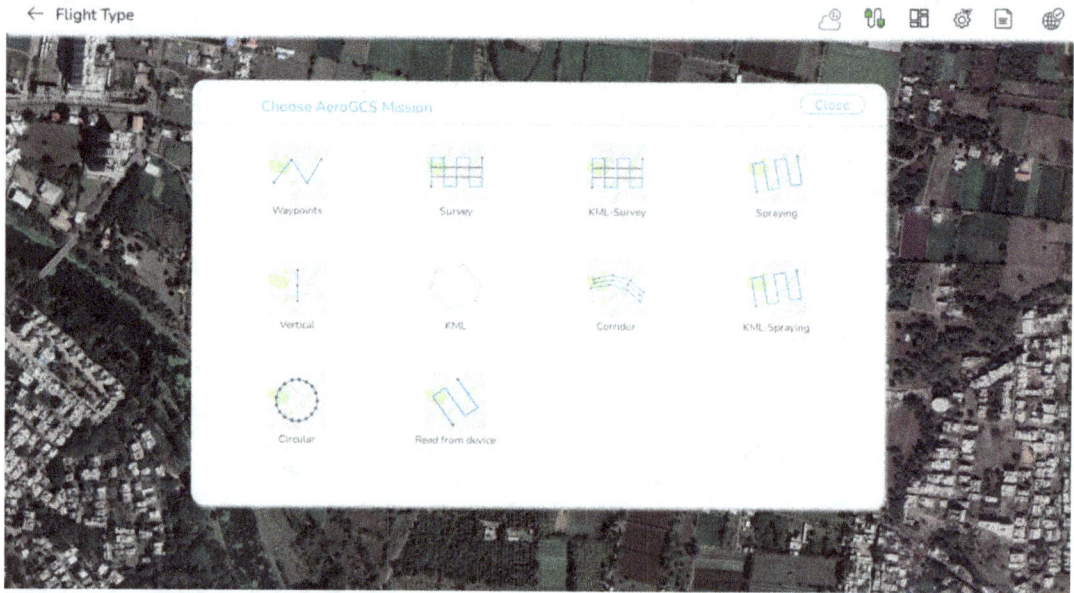

Figure 7.20 – Different flight paths

Out of the 10 different types of mission planning that AeroGCS offers, we will only study a few of them, as these are the ones needed for beginners and form the basis of other kinds of waypoint planning, which are suitable for different applications.

Waypoint planning

A waypoint is an imaginary point on the surface of the Earth that is represented on a map, which includes some typical information such as latitude and longitude. These points are helpful in making a complete flight path through which a drone can travel. So basically, the drone follows the information in the waypoint by using its in-built GPS.

The following screenshot is a flight-planning window from AeroGCS that offers different waypoints and decides on the altitude uploaded to the drone. The compass on the top-left corner shows the direction of the drone (e.g., where the drone is heading), and the window on the right side shows the characteristics of the waypoint, such as latitude, longitude, altitude, and speed. This also gives the total distance that has been traveled by the drone following all the waypoints:

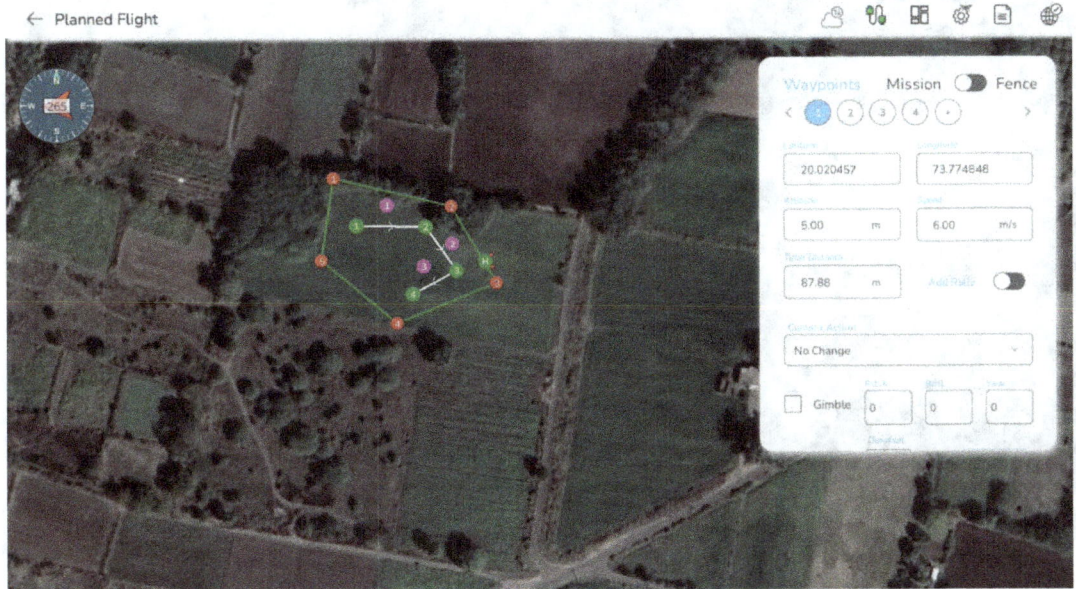

Figure 7.21 – Waypoint mission plan for geofence

How to plan a mission

Waypoint machine planning is simple using the following steps:

1. Click the points on the maps where you want the drone to go to create a flight path. Make sure that the flight path does not collide with any of the obstacles on the ground in real time.
2. Decide the altitude of each waypoint based on the screen and the speed at which the drone should fly.
3. A gimbal is used to take photos and video using a camera, which can be controlled by the software.
4. Save the mission you have developed on the drone.

Survey planning

Survey planning, as the name suggests, refers to a type of mission planning that is used to conduct aerial surveys of huge land masses, where the drone follows specific grid-like patterns and flies at an altitude of 100-120 m above ground. The flying drone has a camera as a payload, and it takes images of the area at certain intervals. These images are geotagged and are later used to create a map of the complete area. Since surveying is an important industrial application, these missions come preloaded and can be configured by adjusting some of the parameters for a fruitful result. The following screenshot shows the typical view of a survey-planning mission:

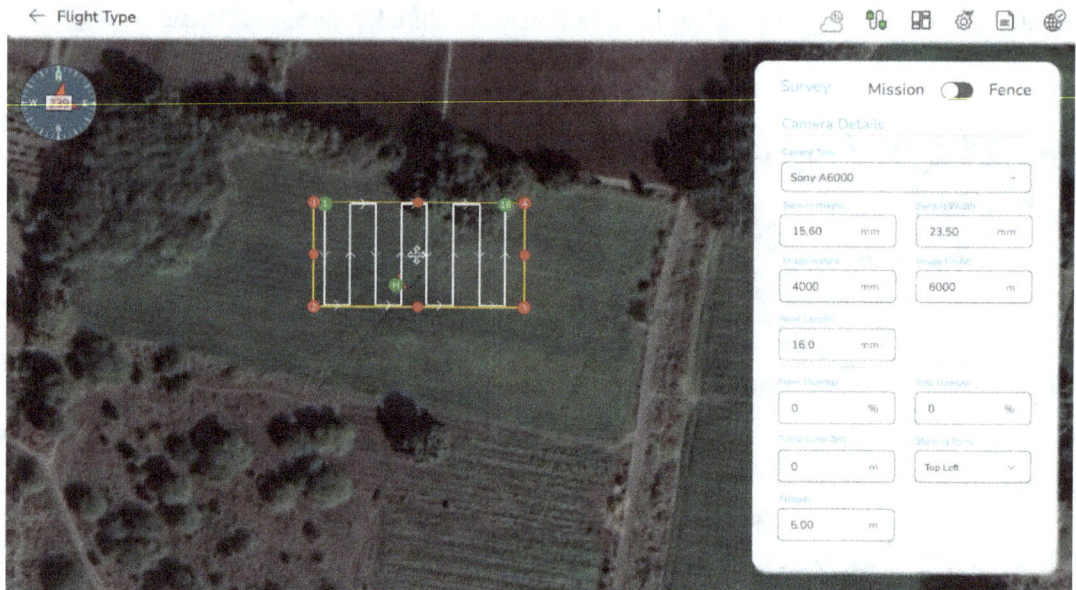

Figure 7.22 – A survey mission with flight parameters

We have reached the end of this chapter. We have studied the basics of the GCS software that is available. We have covered in depth how to configure a drone using AeroGCS. This gave us a brief understanding of and familiarization with the working of this type of software. There is more to understand regarding GCS software, which can be enhanced with practice.

Summary

In this chapter, we provided an overview of GCS software, including what it looks like and what its functionalities are. We also saw the various kinds of options it offers in terms of mission planning and the various kinds of flight paths we can plan.

We also learned about the methods that can be used to connect a drone to the GCS software. However, we won't be stopping here.

In the next chapter, we will use this GCS software to perform other actions, such as calibrating sensors and actually flying a drone.

8
Understanding Flight Modes and Mission Planning

Previously in this book, we explored various **Ground Control Station** (**GCS**) operations and functions, as well as how they help to command and control a drone in real time. We also saw how a GCS is used to configure a drone and its various peripheral devices through its numerous interfaces. In previous chapters, we learned how to configure a drone so that it can use different flight modes that have already been configured by the respective developers from ArduPilot and PX4. In this chapter, we will see what those flight modes are, how they help a pilot fly the drone, and what sensors they use. This chapter will help you to understand the different kinds of flight modes under which a drone flies and how these flight modes are activated using a remote control or GCS. This will, in turn, help you understand more about flying a drone using a remote control or GCS.

In this chapter, we are going to cover the following topics as they will help us understand each flight mode and how to set it up:

- Flight modes and types
- Assigning flight modes via remote control
- Assigning and changing flight modes in a GCS
- Planning and executing an autonomous waypoint mission

Technical requirements

For this chapter, you will need to understand the working principles of IMU sensors and how they work. You will also require knowledge of basic physics and gadget-handling skills. Apart from this, you will need to go through the documentation for pre-existing flight stacks such as ArduPilot and PX4 since we'll discuss the basics which are developed in these two flight stacks.

Flight modes and types

Flight modes are predefined settings that govern how an autopilot system dictates the drone's behavior during the flight. They also define which sensors are to be used and how much manual intervention is allowed by the pilot. In summary, flight mode is a set of instructions that help the autopilot system behave in a predefined manner while flying.

We can set flight mode using a GCS or directly via a remote control; thus, it's easy for the pilot to switch between two flight modes depending on the situation.

In general terms, we can say that flight modes assist the pilot in flying the drone more easily. They allow certain parameters to be controlled, such as altitude and position, which help the pilot to fly the drone smoothly since autopilot takes control of the drone based on its sensor readings, as well as the response the pilot provides as input. The term flight mode is common in various unmanned vehicles, such as multi-copters, fixed-wing aircraft, and ground vehicles. All vehicles respond differently in different kinds of flight modes based on their physical structure and what they are expected to do.

In the next few sections, we'll discuss what the flight modes are and how we can set them up using a remote control or a GCS. Setting the flight mode is the most important part when setting up a vehicle for flying because it tells the vehicle how the pilot is flying the drone and how it should respond.

Types of fight modes

As we discussed previously, multiple types of flight mode switches can be implemented on a multi-copter for smooth and easy flying. These flight modes depend on the type of unmanned system, as well as its configuration. For the sake of this section, we will only study the modes that are used in a multi-copter while considering the prominently used ArduPilot and PX4 flight stack.

Based on the inputs used by the flight controller, flight modes can be divided into three categories: **manual**, **semi-autonomous**, and **autonomous**. Each mode requires some of the onboard sensors to be used by the flight controller, such as GPS and barometer. For each flight mode, there has to be an input to the flight controller, via either the onboard sensors or an external source, such as a remote control. Sometimes, they use both the onboard sensor and the remote control to maintain their level and complete the instructions given in the desired flight mode. Some examples of flight modes include manual, position, altitude, and stabilize.

These flight modes are also categorized based on their level, from beginner to expert. A beginner pilot learning the stick controls on a remote control would be provided semi-autonomous modes where the drone can level itself in the air and the beginner pilot can only control its forward, backward, and sideward movements – that is, roll, pitch, and yaw – which would help the pilot understand the basics of drone flying. The pilot can then move on to using some expert modes, such as autonomous and manual. Some flight modes are used in high-speed flying, such as manual and acrobatic, to make the drone do stunts and flips. Let's look at the different kinds of flight modes that are available and can help pilots. For a beginner pilot, manual aspects such as takeoff and landing can also be achieved autonomously with the help of automatic commands, which are served by the GCS.

With that, we've seen the importance of flight modes. They are the primary way to set up the drone's flight. Now, let's learn about different kinds of flight modes and how to set them up.

Manual mode/stabilize mode

Manual mode is a flight mode where the pilot has complete control over the drone and has no involvement in the flight controller. The drone's movement is in the hands of the pilot sticks, where the drone moves based on the inputs given by the pilot, hence the name **manual mode** or **stabilize mode**. While flying in this mode, the autopilot system does not receive any input from the sensor. The pilot must be able to see the drone and fly it. This is an advanced mode of flying that requires good experience and muscle memory over the remote controller to stabilize the drone in the air and fly it correctly.

We call this mode complete remote controller mode when the centered roll and pitch sticks level the vehicle but the position of the vehicle has to be managed manually using the roll pitch stick and altitude (up and down) using the throttle stick. The yaw is managed by the left and right rotation of the throttle stick.

The roll and pitch sticks also control the angle of the vehicle – that is, the **attitude**. As soon as we release the control sticks, they will return to their central point and the vehicle will keep hovering in place. It will be mechanically balanced and maintain its altitude until no external force is applied.

The following figure shows the manual flight mode and stabilization implementation in a remote controller, as well as the behavior of the drone. A remote controller typically consists of many switches and potentiometers, which help us control and configure the peripherals that are attached to the drone:

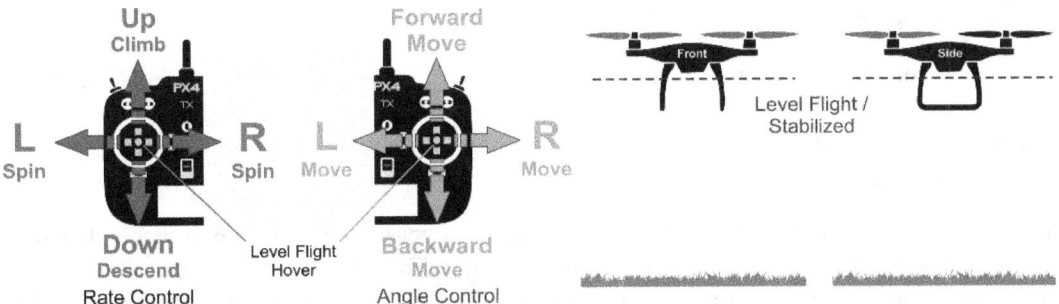

Figure 8.1 – Manual mode flight

Semi-autonomous modes

As discussed previously, semi-autonomous modes are flight modes where partial inputs are given by the flight controller to level the drone up using its major sensors, but major inputs, such as altitude and roll and pitch moments, are given by the pilots. These are also called assisted modes as the pilot is being assisted by the flight controllers and sensors to ensure smoothness in terms of position and altitude. Let's take a look at some of the assistant modes that are available.

Position mode

Position mode is an easy flight mode and is helpful for beginners when it comes to flying a multi-copter. In this mode, the roll and pitch sticks control the acceleration over the ground and make the vehicle move forward, backward, left, and right, while the throttle sticks control the up-and-down movement of the drone. When you release the remote controller sticks, the drone comes to a halt – that is, the vehicle stops accelerating and remains in position. Since it is an assisted mode, the flight controller makes use of the GPS to lock the position where the drone is currently flying and continuously holds that position, despite different forces, such as wind. It is the safest flying mode for beginners:

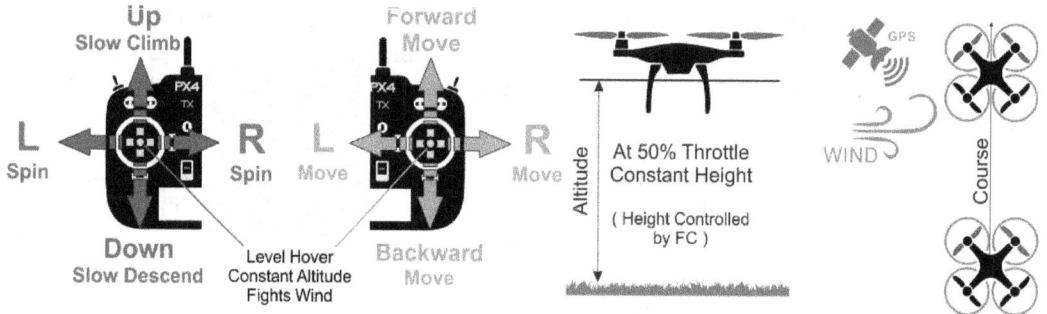

Figure 8.2 – A drone in position mode

Attitude hold mode

Attitude hold mode is also a relatively easy flying remote controller mode in which the roll and pitch sticks control the moment in the left-right and forward-backward directions, the yaw stick controls the clockwise and anti-clockwise movement of the drone, and the throttle sticks control how the drone ascends and descends.

When these sticks are released, they acquire their center positions on the remote controller and hence the drone will also be leveled and maintain its current altitude. The important point here is that unlike position mode, the drone will not maintain its position and instead will maintain its altitude. The flight controller will make use of its barometer sensor or the altimeter sensor (if installed) to maintain the current altitude decided by the pilot.

Hence, we can say that in this mode, the drone will not maintain its position over the ground, unlike in position mode. The pilot has to manually control the drone using the roll and pitch sticks to control its position. However, the flight controller would control its altitude, which would be fixed by the pilot:

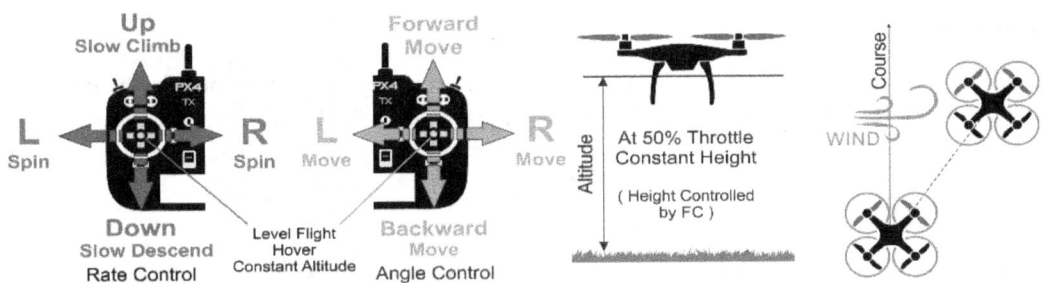

Figure 8.3 – A drone in attitude hold mode

Autonomous modes

Autonomous modes are flight modes that do not require a pilot's input through the remote controller; rather, these modes are completely autonomous and can be planned through the GCS. Let's consider the different modes that can be switched through a GCS or a remote controller by assigning a mode switch.

Hold mode

As its name suggests, hold mode helps hold the drone at its current GPS position and respective altitude without the need for the pilot's input. This assists in leveling the drone. In this mode, the flight controller receives help from the GPS sensor and the barometer sensor to lock the drone at a particular GPS coordinate and attitude. Due to this, the drone can maintain its position against the wind or any other forces. This mode can be used to pause a mission and regain control over the drone in the case of an emergency. This mode can also be configured via the GCS or be assigned to any switch on the remote control to activate it.

Return mode

Return mode is the most important recovery mode in the drone as it helps the pilot get the drone back to its original take-off position. This helps the drone fly a clear path to a safe location and it can be activated manually, by remote control, or through a failsafe. This mode can be configured to activate during any failsafe condition, such as low battery, data link loss, or a geofence breach. The behavior in this mode can be configured and decided by the GCS before flight, depending on the environment's conditions. This mode makes use of the GPS location and hence guides the drone to that particular location by itself before returning to its original take-off position, hence its name.

Mission mode

As its name suggests, in mission mode, the drone executes a predefined mission that is programmed by the GCS. While programming a mission, the pilot can give the drone different waypoints at different altitudes with a desired speed. After switching to mission mode, the drone goes through the various waypoints at different altitudes. In this mode, the drone makes use of all its IMU sensors and additional peripheral devices before returning.

Take-off mode

Take-off mode helps us to do an automatic takeoff without using a remote controller by giving a suitable altitude to the drone via the GCS. Upon reaching that altitude automatically, the drone comes to a hold position. At this point, it holds its altitude and its GPS location and continues hovering.

Land mode

This mode causes the multi-copter to land automatically without any input from the pilot. This mode can be configured at the end of any mission or upon triggering any failsafe conditions.

Assigning flight modes via remote control

So far, we've learned that the remote controller is a key element of flying drones in manual mode, where the pilot controls them. As we have seen, a remote controller typically consists of many switches and potentiometers, all of which help us control and configure the peripherals that are attached to the drone. With the help of these switches and potentiometers, a pilot has full control of the drone and can switch from one mode to others with the help of switches. In this section, while using a remote controller, we will use the same MAVLink protocol on different channels to configure the various flight modes.

In a remote controller, a channel is an independent communication pathway between the transmitter and the receiver on the drone. Each channel is assigned to a specific function or control, such as throttle, pitch, roll, or yaw, allowing the pilot to manipulate different aspects of the drone's movement and features.

Nowadays, there are many remote controllers on the market, such as FrSky, Flysky, and SIYI, and some of them have an integrated GCS. Setting up switches and flight modes is common in all kinds of remote controllers. We advise you to read the user manual of the remote controller that you are using as this will help you go through the following steps and perform them. Here, we will be using a generic methodology to enable the switch and assign it to a channel. This may vary from controller to controller.

1. Enabling the switch on a remote controller: Every remote controller has certain switches and potentiometers that are not enabled by default, which means they need to be set up. It is always advised to select three ways to switch on the remote controller as this will help you set up three different flight modes on a single switch. How the switch will be enabled will be different for different remote controllers, but once the switch is activated, we need to assign a proper channel on which the switch will work.

2. Assigning a flight mode channel: For a switch to send a signal to the drone and the drone to understand the command given to it, both must be on a common channel. So, the channel that's been set on the switch should be configured via the GCS and the flight mode, as we'll see later. We'll assign channel 5 for the flight mode to the switch. Please make sure that you assign the same channel to the GCS while configuring the flight mode.

3. Configuring the flight mode in the GCS and setting the switch: Later in this chapter, we will learn how we can configure the flight mode in the GCS according to the settings in the remote controller. However, in this step, we'll set up different flight modes for our switch. When you move this switch, you will see that the indicator goes to a different flight mode; this is something that can be configured on the flight screen. You can also choose the type of flight mode you want to assign to a switch. This means that three flight modes can be assigned to one switch.

With that, we've learned how to configure flight mode on a remote controller, which helps us use different flight modes while the drone is in the air. We can also switch to another flight mode that's not been configured on the remote controller via the GCS. Similarly, the remote controller can be switched to different channels; with those different channels, we can attach different peripherals, such as a gimbal, sprayer, or camera, that we want to control with the help of the remote controller.

How it works

We've broken the **pulse width modulation** (**PWM**) into different fragments, where each fragment can be used for a specific function in the same channel. We now know that a PWM channel works between 1,000 and 2,000 microseconds, with 1,000 being the lowest and 2,000 being the highest in terms of a signal's intensity. When we enable this PWM on the three-way switch, the limit gets divided into three equal parts, which means that the PWM is divided into three fragments, where each fragment sends different commands but on the same channel we have set. Thankfully, we can send different messages via the same channel by breaking it into segments with the help of a switch. This also helps us control the various peripherals with fewer resources.

Assigning and changing flight modes in a GCS

In this section, we'll take a deeper dive into *step 3* from the previous section. Here, we'll assign the three different flight modes to the switch that we configured previously. As we know, a GCS is the prime source of configuration for a drone, so we'll use this knowledge to configure the flight mode that we want to put on this switch so that we can change it while it's flying.

Flight mode configuration is also a key requirement that is of utmost importance before flying the drone and without which the drone could not be flown with the help of a remote controller. Here, we will see how we can configure flight mode in the GCS and how we can change the flight mode through the GCS.

Configuring the flight modes of the remote controller switch

When we open AeroGCS, we'll see that the various flight modes are controlled through the radio (via a transmitter switch) or using commands from AeroGCS. Select the flight mode in which you're comfortable flying and that you wish to assign to the switch.

1. **Connecting to the drone:** Connect the GCS to the drone via a USB cable or telemetry and let the parameters be downloaded.
2. Navigating to **RPA Configuration | Flight Modes**: Here, you will see a window similar to what's shown in the following screenshot. It will tell you about the flight mode channel and the different flight modes that you can select from the drop-down menus provided:

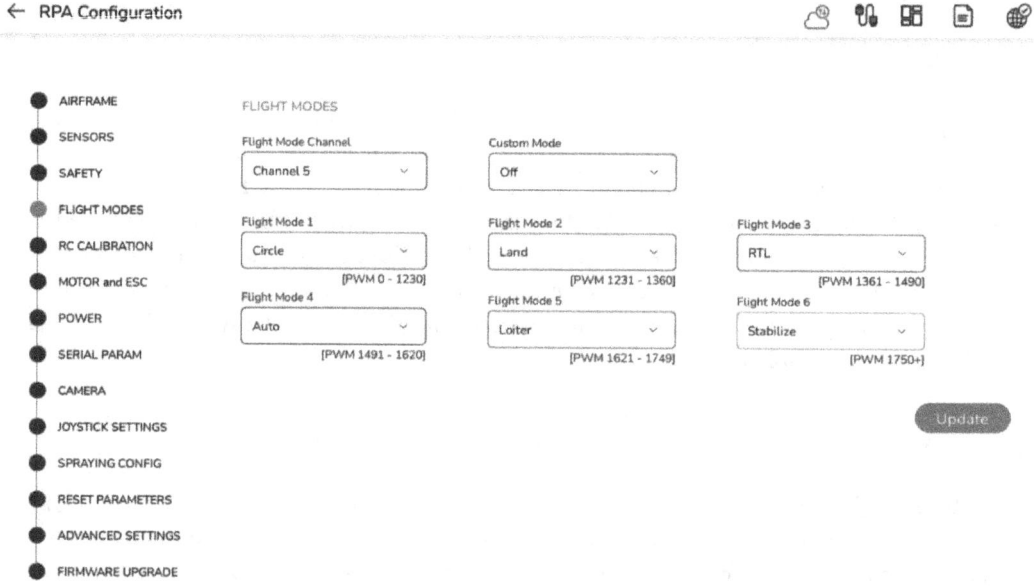

Figure 8.4 – The Flight Modes window in AeroGCS

In the **Flight Mode Channel** field, select the same channel you chose to assign the switch in the RC. In our case, we are going ahead with **Channel 5** since we assigned that to the switch previously.

As shown in the preceding screenshot, you can choose from various flight modes, from flight mode 1 to flight mode 6. Below these options, you can see the PWM values for each. We learned about these PWM value ranges in the previous sections. Here, different flight modes have been assigned a specific range of the PWM, which clarifies which flight mode is selected at the moment. A similar kind of range and segmentation is being done on the remote controller site, where we assign a channel to the three position switches. The PWM is segmented into three different ranges, which indicates which PWM has been selected; the same is reflected in this window of the GCS:

Flight Mode	PWM Range
Flight Mode 1	0 – 1230
Flight Mode 2	1231 – 2360
Flight Mode 3	1361 – 1490
Flight Mode 4	1491 – 1620
Flight Mode 5	1621 – 1749
Flight Mode 6	1750 +

Figure 8.5 – Segmentation of PWM

3. Moving the switch and selecting a flight mode on the GCS: Upon selecting **Channel 5**, the various flight mode options will appear. This indicates that this flight mode is being chosen by the position of the switch. If you move your switch to a different position, a different flight mode will be highlighted; again, if you move your switch to a third position, a new flight mode will be highlighted.

Whichever mode is highlighted at the position of the switch indicates the PWM range chosen by the switch. This will also help you select the desired flight mode for all three positions of the switch. This process is depicted in the following screenshot. As you can see, there's also a drop-down menu where you can choose the desired flight mode and position of the switch:

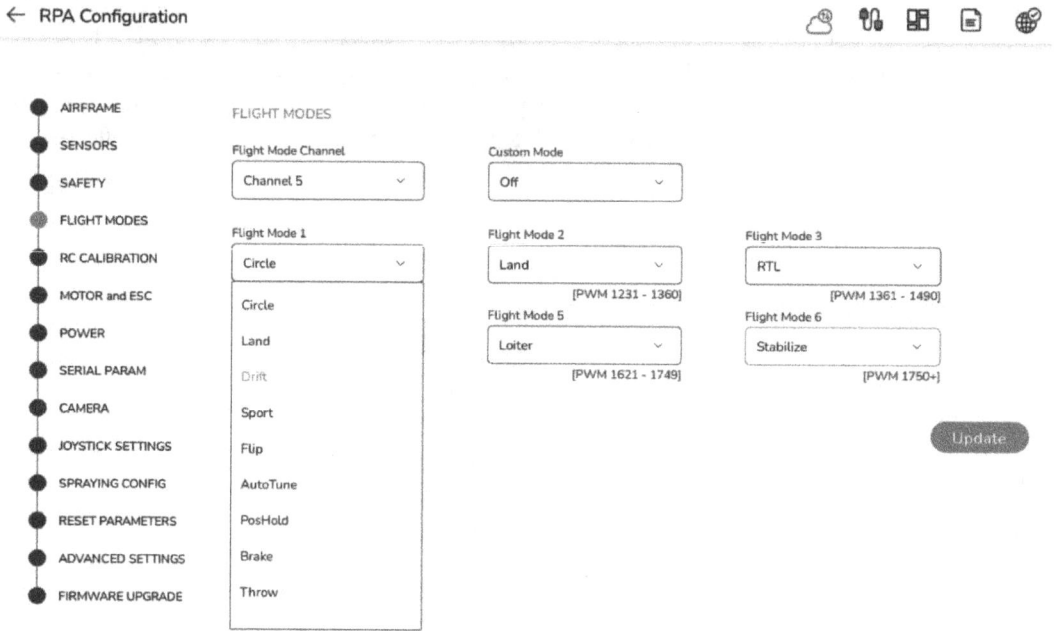

Figure 8.6 – Selecting different flight modes

As you can see, **Flight Mode 6** is highlighted and the selected flight mode is **Stabilize**. This flight mode can be changed at your convenience using the drop-down menu for that flight mode.

4. Saving the configuration: Click on **Update** and save the flight mode so that it can be used. You can select a different position and check whether the flight mode changes in the dashboard, as shown in the following screenshot:

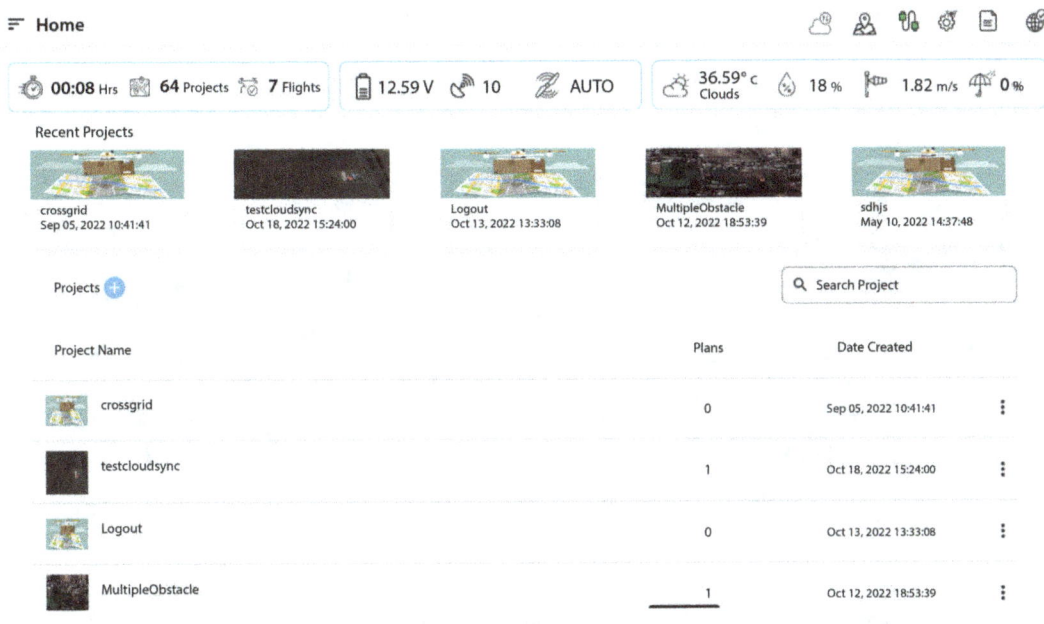

Figure 8.7 – Home screen of AeroGCS

In this section, we learned about different flight modes and how to set them up using a GCS and a remote controller. Next, we'll learn how to create and execute an autonomous mission with the help of a GCS without using a remote controller. However, there is always a remote controller command that can override the GCS command while executing a mission so that the pilot can take control of the system in an emergency.

Planning and executing an autonomous waypoint mission

Previously, we looked at different kinds of flight modes, one of the most important of which was autonomous mode. Flying in autonomous mode requires creating a mission plan, something that's done in the GCS software and then uploaded to the flight controller through a data link. Let's see how can we plan a mission, upload it to the drone, and execute it.

1. Logging in to your profile : Use your email and password to log in to your AeroGCS account. Note that not every GCS requires this step; some will take you to the dashboard. However, in this case, the GCS has been customized so that each user can handle their own projects:

158   Understanding Flight Modes and Mission Planning

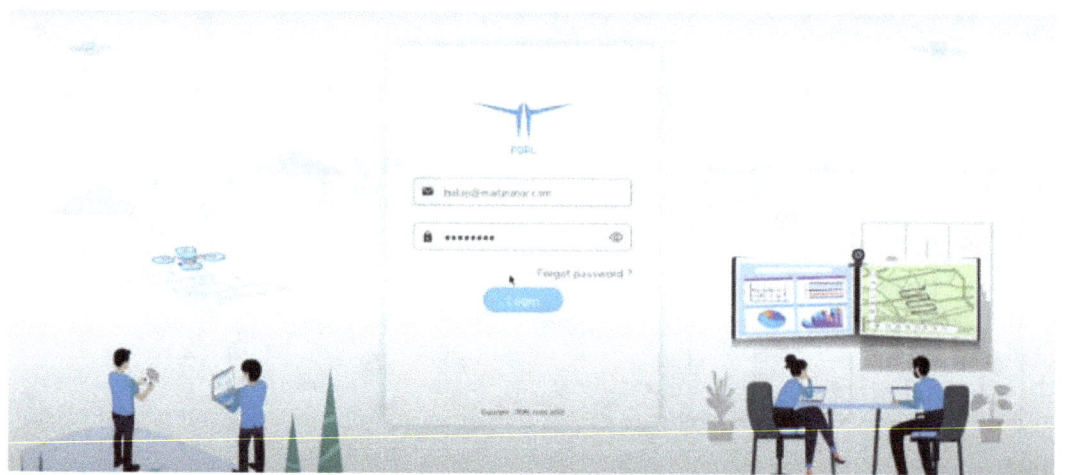

Figure 8.8 – Login window of AeroGCS

2. Connecting the GCS to the drone: Before we do anything else, we must connect the device. Click on **Yes** and connect the device using the option given in the second window:

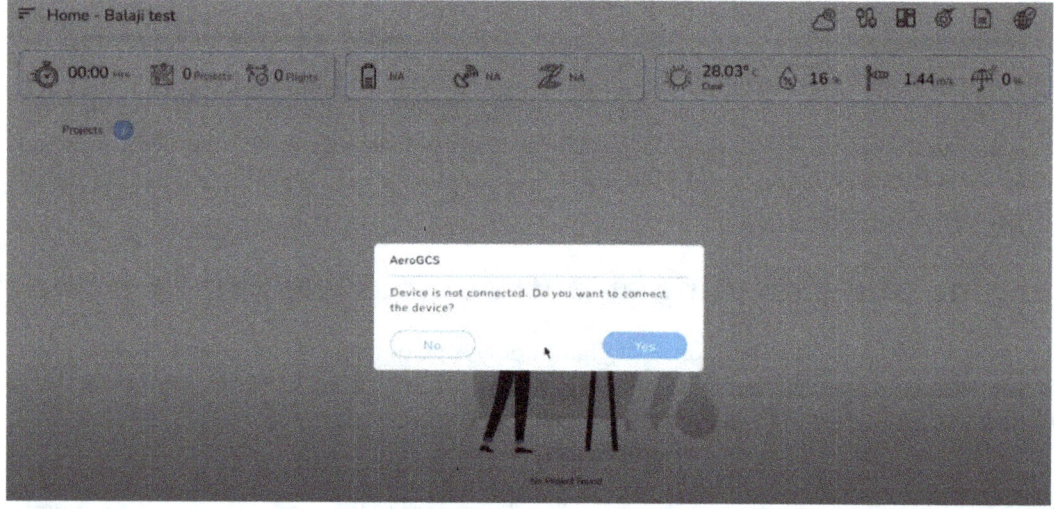

Figure 8.9 – Connection protocol

As shown in the following screenshot, we have different ways to connect to the drone – that is, **Serial**, **TCP**, **UDP**, and **Bluetooth**. We'll go with a serial telemetry connection to the drone here, set the baud rate to 115,200, and proceed:

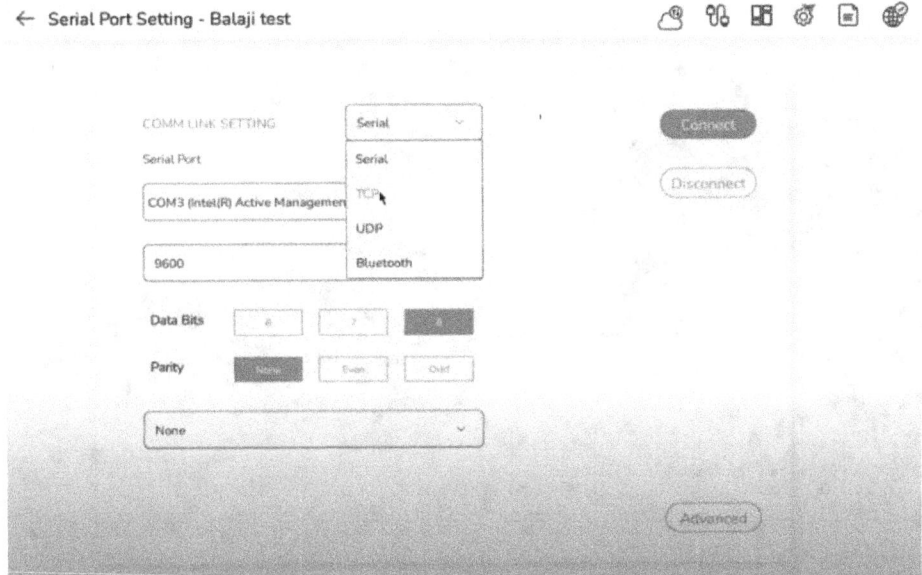

Figure 8.10 – Selecting a connection protocol

3. Setting a project name: Setting a project name will help us create customized projects for the client and also save projects that we have completed in the past so that we can repeat them as required:

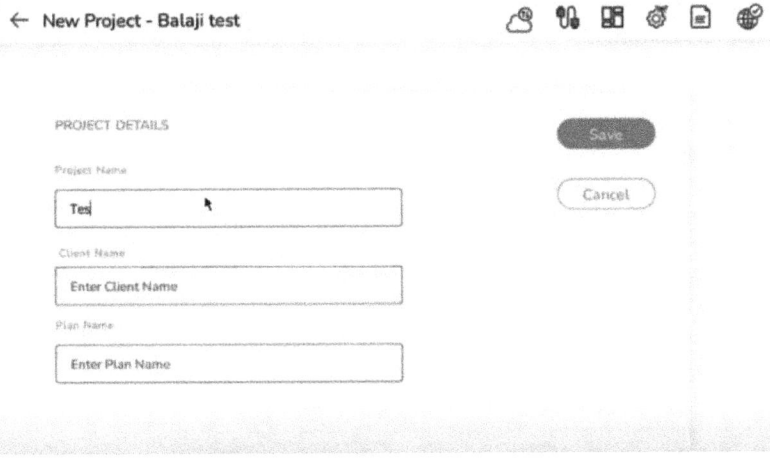

Figure 8.11 – Setting a project name

4. Selecting a flight location: With the help of the drone's GPS, you will be able to automatically detect its current location and see a map of where the mission has to be planned. Click **Next** if the location that's been chosen is correct based on the environment in which you're currently present:

Figure 8.12 – Location selection in AeroGCS

5. Choosing a flight plan: Based on the application's requirements, AeroGCS offers customized plans that can be modified as per your requirements. The user can use any of the flight plans they want to perform a specific task, such as survey, spray, or inspect. Though this feature is present in AeroGCS, in some other GCSs, the plan must be created manually. We'll select a waypoint mission here as we are going ahead with an autonomous waypoint mission:

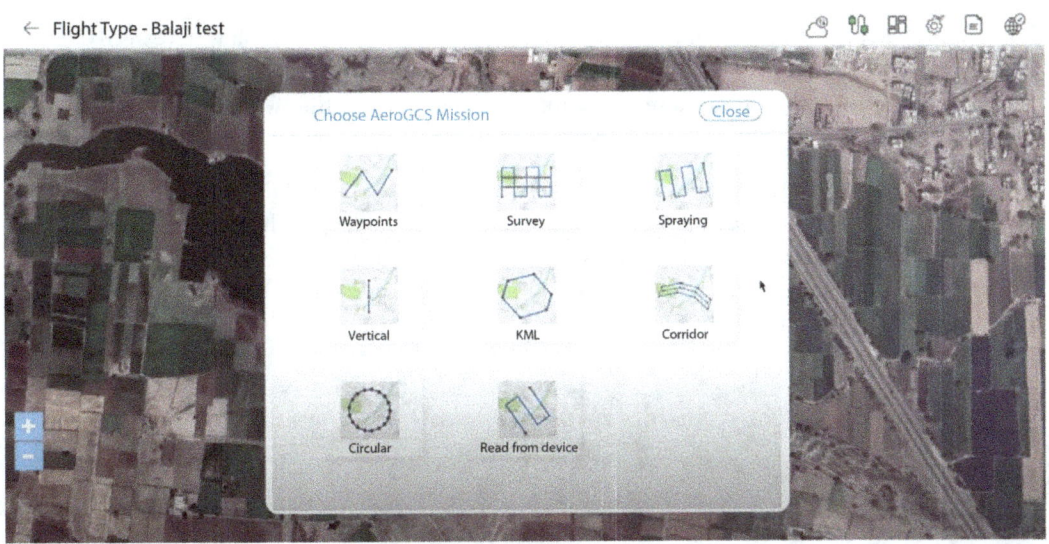

Figure 8.13 – Available flight plans

6. **Specifying the flight parameters**: Now, you need to specify the flight parameters for the drone. We need to specify the altitude at which we want the drone to fly, as well as its speed. These parameters will help us decide on the behavior of the drone and can be changed as per the application's requirements. Once you've done this, click **Next**:

Figure 8.14 – Choosing a flight type

7. **Creating a geofence:** As shown in the following screenshot, various waypoints have been created and a flight path has been automatically generated on which the drone will fly. Now, we need to create a geofence around these waypoints so that the drone flies within the four points of the geofence.

   A geofence acts like fencing around the drone and is created by certain points that define the area within which the drone should fly. Crossing these would trigger the failsafe action and the drone will land or return, as configured in the failsafe action. The geofence option prevents the drone from flying away in an emergency and restricts it to a limited area:

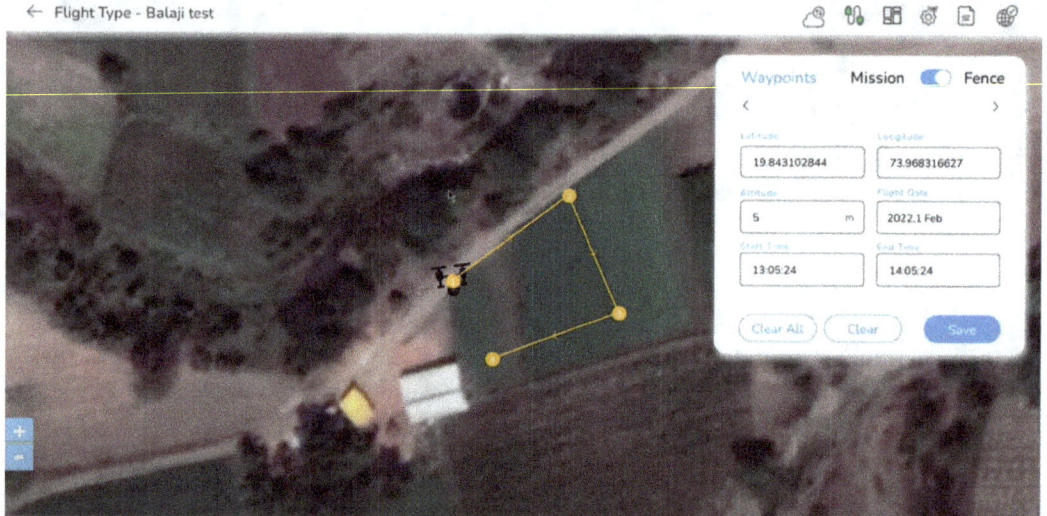

Figure 8.15 – Waypoint mission created

As shown in the following screenshot, the geofence has been created around the mission and is indicated in red.

Click **Save** to program the mission into the drone:

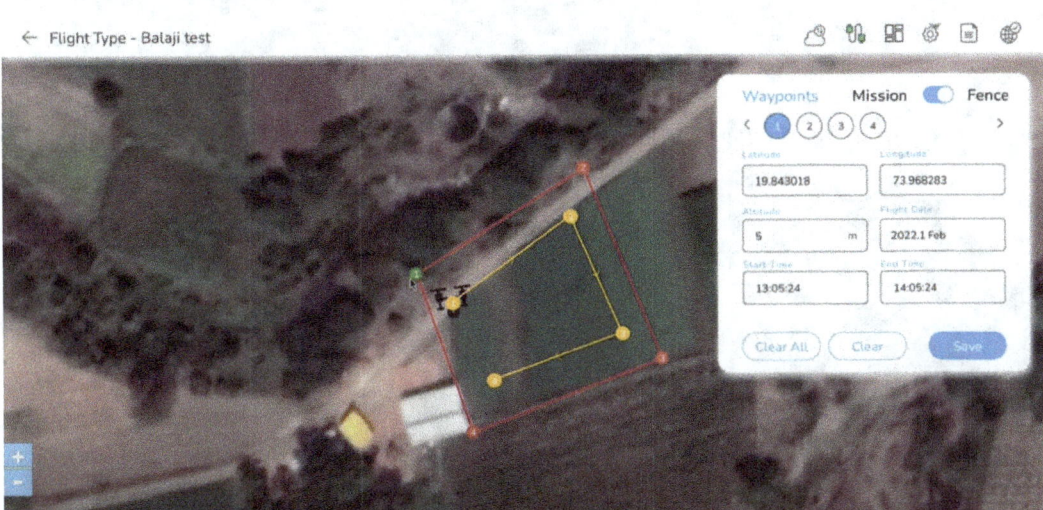

Figure 8.16 – Geofence created

8. Take-off and pre-flight checks: Click the **Take Off** button on the left-hand side to start the mission; a pre-flight check window will pop up. This window will recheck everything before the final takeoff. This also helps the pilot reconfigure every setting and see the system's health before the drone takes off and starts its mission:

Figure 8.17 – Mission start window

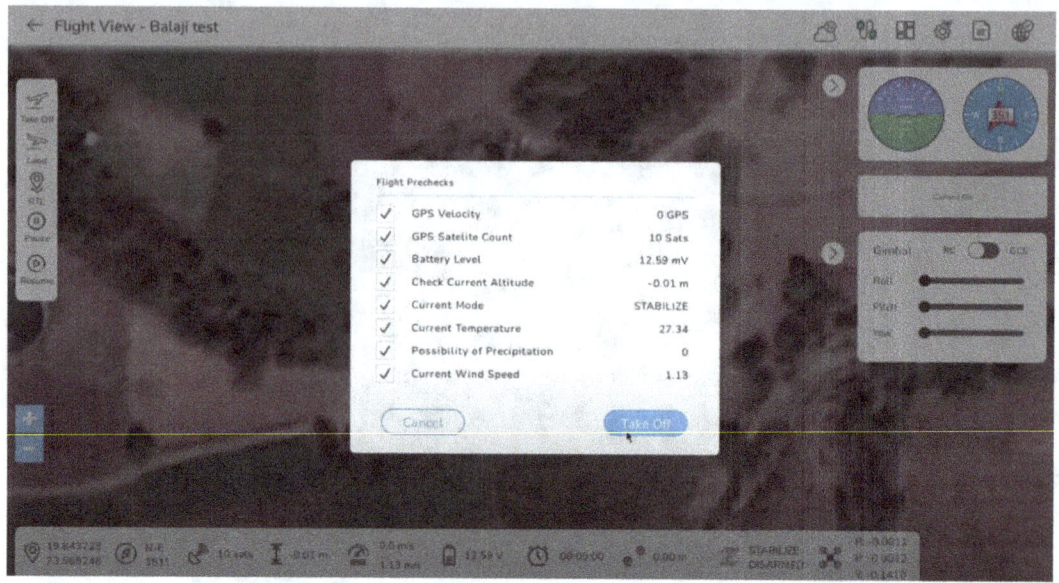

Figure 8.18 – Pre-flight checklist

9. **Executing the mission and landing:** Once you've given the command, the drone will take off and start executing the mission. After completing its mission, it will land at its designated point. All the important parameters, such as battery coordinates and GPS altitude, are reflected at the bottom of the screen, which provides real-time information on the system's health during the mission:

Figure 8.19 – Mission execution in progress

With that, we have successfully created and executed a mission via a GCS. Different kinds of missions can be programmed and executed by a drone. It is recommended that you download the software that was used in this chapter and practice this kind of mission planning regularly so that it becomes a habit.

Summary

In this chapter, we learned about the most important part of a flight – flight modes. This chapter has helped you understand how a simple drone can be flown in different ways and how flight modes help us execute many commercial applications that use drones. We also learned how to use different platforms to execute different projects. Finally, we saw how the manual and autonomous modes work when it comes to flying and how to program a drone so that it can fly in autonomous mode.

In the next chapter, we'll learn how to tune different parameters of the drone so that we know what final checks must be done before a flight.

Part 3: Configuration, Calibrations, Flying, and Log Analysis

This is the final step after our drone is ready for flight. This includes flashing firmware to the system and selecting the right frame. This part will guide you to the manuals of flying and configuring your system to make it ready to fly. It will also help you to analyze flight performance based on the logs recorded in the flight data recorder.

This part has the following chapters:

- *Chapter 9, Drone Assembly, Configuration, and Tuning*
- *Chapter 10, Flight Logs Analysis and PIDs*
- *Chapter 11, Application-Based Drone Development*
- *Chapter 12, Development of Custom Survey Drone*

9

Drone Assembly, Configuration, and Tuning

In this chapter, we'll cover multiple aspects of a drone in terms of its hardware. Previously, we studied the theory behind how a drone hovers and moves in the air. We also studied the different components that are required when assembling a drone and the software that is required to fly and configure one. In this chapter, we'll dive into the fundamental aspects of drone assembly and configuration, offering insights into the intricate components that make up a drone, the tools required for building one, and the software and firmware adjustments necessary to optimize its performance. From selecting the right frame, motors, and flight controllers to calibrating sensors and setting up remote controls, this guide will provide you with the knowledge and guidance needed to embark on your drone-building journey before its maiden flight.

This chapter will help us to understand the end-to-end process of assembling a drone and making it mechanically stable in all X, Y, Z directions so that the total force of the drone is directly pointed downward, perpendicular to the ground.

To understand things in depth and apply them to real life, we'll cover the following topics as they will help us learn about the skills, tools, and precautions for building a drone from scratch:

- Components and tools required to build a drone
- Assembling the components of a drone
- Setting up and configuring avionics
- Understanding calibration and failsafes
- Setting up a maiden flight

Technical requirements

Building and flying a drone requires a combination of technical skills and software knowledge. We studied these skills and components in the previous chapters and they will help you here. Let's take a closer look:

Technical understanding of components:

- **Knowledge of drone components**: You need to understand the various components of a drone, including its frame, motors, **electronic speed controllers** (**ESCs**), flight controller, battery, and sensors
- **Understanding of power systems**: You need to be familiar with **Lithium-Polymer** (**LiPo**) batteries, including their voltage and capacity, and how to calculate power requirements for your drone
- **Understanding of propellers**: You need to know how to choose the right propellers based on your motor specifications and intended use

Electronics and wiring skills:

- **Soldering**: You must be proficient in soldering to connect wires, components, and connectors securely
- **Circuit understanding**: You must understand basic electronics and circuitry so that you can troubleshoot and repair electrical issues

Programming and software skills:

- **Flight controller configuration**: You must know how to configure and program the flight controller using software such as Betaflight or Cleanflight, or even software provided by the manufacturer
- **Transmitter setup**: You must be able to configure and calibrate your transmitter (remote control) to communicate effectively with the flight controller
- **Understanding of PID tuning**: You must understand how to tune the **Proportional-Integral-Derivative** (**PID**) controllers to optimize your drone's flight stability and performance

Flight skills:

- **Basic flying skills**: You must have proficiency in basic drone flight maneuvers, including takeoff, landing, hovering, and basic navigation

These technical skills are necessary as you need a physical and analytical method to build the drone and have strong wiring and component skills.

Components and tools required to build a drone

To begin the process of assembling a drone, we must focus on gathering the essential components we will need. These include the drone's frame, motors, ESCs, propellers, flight controller, battery, and transmitter/receiver. Before we dive into the intricate assembly steps, having these components at hand is the crucial first step. So, let's have a look at the complete list of the materials that are required for assembling the drone.

In this section, we will study the different tools and components that are required to assemble a drone. We will only focus on how to assemble the different components as we have already studied the various sensors, physics, and dynamics of the drone.

The complete list of build materials and tools, along with the **bill of materials** (**BOM**) of the drone, are provided in the following tables:

Tools:

Tool	Quantity	Purpose
Screwdriver set	1	To assemble and secure components
Soldering iron	1	To solder electronic connections
Soldering wire	1 roll	Provides a conductive path during soldering
Wire stripper	1	Removes the insulation from wires
Heat shrinks tubing	Assorted	Insulates and protects soldered connections
Pliers	1	To grip and bend wires
Hex wrench set	1	To tighten hex screws on the frame and motors
Multimeter	1	To measure voltage, current, and resistance
Tweezers	1	Precise handling of small components
Nut driver set	1	To tighten nuts on bolts and screws
Wire cutter	1	To cut wires cleanly
Safety glasses	1	Eye protection while soldering
Anti-static mat	1	To protect electronic components from static
Anti-static wrist strap	1	To ground static electricity from the assembler

Table 9.1: List of tools required

BOM:

Component	Quantity	Specifications
Frame	1	Carbon fiber, with sufficient payload capacity
Flight Controller	1	GPS capabilities, compatible with survey applications

Motors	4	High torque, brushless motors
Electronic Speed Controllers	4	Matched to the motors, with sufficient current rating
Propellers	4 pairs	High-efficiency, compatible with the motors
Battery	1	High-capacity LiPo battery, suitable for long flight times
Radio Transmitter and Receiver	1 set	Long-range, reliable communication
Power Distribution Board	1	Distributes power from the battery to various components
GPS Module	1	Provides accurate positioning data for surveying
Camera Gimbal (optional)	1	Stabilizes the camera for clear survey images
Camera (optional)	1	High-resolution camera with survey-specific features
Telemetry System	1	Enables real-time data transmission between drone and user
Spirit leveler	1 set	Checks the level of the motors, propeller, and airframe
Propeller Balancer	1	Balances propellers for smoother operation
Power module	1	6s Digital

Table 9.2: BOM for a quadcopter

Wires and connectors:

Item	Quantity	Specifications
Silicone Wire (12-22 AWG) Red and Black	Assorted	High-quality, flexible wire for various connections
XT60 and XT90 Connectors	Assorted	Connectors for the battery and power distribution board
JST Connectors	Assorted	Connectors for smaller electronics and sensors
Servo Extension Cables	Assorted	Extension cables for connecting components at a distance
Heat Shrink Tubing	Assorted	Various sizes for insulating and protecting soldered wires

Table 9.3: Wires and connectors

Assembling the components of a drone

In this section, we will learn how to assemble the various components of the drone. To do this, we will mount them all with fasteners.

There is always an assembly manual that's given by the manufacturer to assemble the drone's frame. Let's look at assembling the frame of the famous Quadcopter Tarot X4:

Figure 9.1: Tarot airframe

Airframe assembly

Assembling the frame of a drone is a crucial step in building an **unmanned aerial vehicle** (**UAV**). Proper assembly ensures structural integrity and stability during flight. Considering the aforementioned Tarot X4 airframe, this section will provide a precautionary guide you can follow while assembling the airframe to verify its correctness.

Assembling the drone's frame

Considering the famous beginner DIY build for the Tarot X4, we will start by assembling the critical components as per the manufacturer's assembly guide. The guide will be different from airframe to airframe, but the potential steps mentioned here will remain the same.

First, we must identify the frame's components. We familiarized ourselves with the different parts of the drone's frame, including the central frame, arms, and connecting plates, in *Chapter 1*.

Check out the manufacturer's assembly instructions or diagrams and details of fasteners for the airframe you have purchased. Now, follow these steps:

1. **Arm attachment**:

    If your drone frame has removable arms, attach them to the central frame.

    Align the screw holes on the arms with those on the central frame.

    Use the provided screws or bolts to secure the arms in place.

    Tighten the screws evenly to ensure a secure fit but avoid over-tightening to prevent damaging the frame.

2. **Square and level frame**:

    Use a level or a square to ensure that the frame is square and level.

    Proper alignment is critical for stable flight, so take your time with this step.

3. **Check for interference**:

    Verify that no components or wires interfere with the arms' movement or the flight controller's installation points.

    Make sure there's ample space for electronic components, such as the flight controller and **power distribution board (PDB)**.

4. **Secure any additional components**:

    If your drone frame has mounting points for additional components, such as GPS modules or cameras, attach them according to the manufacturer's instructions.

    Ensure that these components are fastened securely.

5. **Visual inspection**:

    Conduct a visual inspection of the assembled frame to ensure that all the components are attached correctly.

    Look for any loose screws, misaligned arms, or damaged parts.

    Confirm that all frame components are secure and that there are no missing parts.

6. **Strength test**:

    Gently apply pressure to different parts of the frame to check for any signs of weakness or flexing.

    The frame should be sturdy and not easily deformed.

Motor assembly

Mounting, configuring, and connecting motors to a drone is a fundamental step in building and setting up a multirotor aircraft.

Figure 9.2: The motor mounting process, front and back

Here is a detailed explanation of how to do this in terms of a quadcopter:

1. **Mount the motors**: Before mounting the motors, you must identify the motor locations on your drone frame. In a quadcopter, these are typically labeled as Motor 1, Motor 2, Motor 3, and Motor 4 and are placed on the respective pins on the flight controllers (refer to the flight controller documentation you are using). We have the following settings:

    - Motor 1: Front right (counter-clockwise direction)
    - Motor 2: Rear left (counter-clockwise direction)
    - Motor 3: Front left (clockwise direction)
    - Motor 4: Rear right (clockwise direction)

2. **Secure the motors to the motor mounts**: Attach each motor to its respective motor mount or arm using screws. Make sure they are securely fastened but not overly tightened to avoid damaging the motor.

3. **Position the motor wires**: Orient the motor wires so that they can reach the central part of the frame where the flight controller and PDB are located. Keep the wires neat and avoid excessive twisting or bending:

Figure 9.3: Clean wiring and insulation

4. **Tighten the motor mounts**: Double-check that all the motor mounts are securely fastened to the drone's frame.

   The next couple of steps will be motor connection and wiring.

5. **Identify motor wires**: Each motor has three wires: two for the power (positive and negative) and one for the signal. Identify these wires on each motor.

6. **Connect the motor wires to the ESCs**: Connect the motor wires to their respective ESCs. The order of connection can vary, so consult the ESC's manual or labeling. It's common to follow a sequence such as Yellow-Blue-Red or Blue-Yellow-Red.

7. **Secure connections**: Solder or use bullet connectors to secure the motor wires to the ESCs. Ensure a strong and reliable connection.

8. **Connect the ESCs to the PDB**: Connect the ESCs to the PDB, typically using bullet connectors or soldering. Connect the positive (red) and negative (black) wires to the PDB's corresponding terminals.

9. **Connect the ESC signal wires to the flight controller**: Connect the signal wires from each ESC to the flight controller's motor output pins. Ensure that they are connected to the correct pins, which are typically labeled M1, M2, M3, and M4.

   From here on we start with the configuration.

10. **Motor direction**: Depending on your flight controller and firmware, you may need to configure the motor direction. Most modern flight controllers allow you to reverse motor direction in software. Ensure that motors are spinning in the correct direction to produce thrust.

11. **ESC calibration**: Calibrate the ESCs to ensure they start and stop at the correct throttle range. Follow the calibration procedure provided in the ESC's manual or your flight controller's documentation.

12. Motor numbering for the quadcopter

    Since we are building a quadcopter, we'll connect the motors to the flight controller in the following fashion. Make sure the motor rotates both clockwise and anti-clockwise:

    - **Motor 1:** Front right (counter-clockwise direction)
    - **Motor 2:** Rear left (counter-clockwise direction)
    - **Motor 3:** Front left (clockwise direction)
    - **Motor 4:** Rear right (clockwise direction)

    The flight controller instructs the motors based on the motor's numbering. This is pre-defined in the motor configuration for the flight. Hence, the motor's ESCs must be connected in the right order.

By following these steps and ensuring proper mounting, wiring, and configuration, you can successfully connect and set up motors on your drone. Always refer to the documentation provided with your specific drone components for detailed instructions and safety precautions.

Assembling and configuring the ESCs

Assembling, connecting, and testing ESCs on your drone is a critical step in building a functional UAV. This section will provide a step-by-step guide on how to assemble and connect ESCs to the drone frame and test them. The following figure provides an example of this:

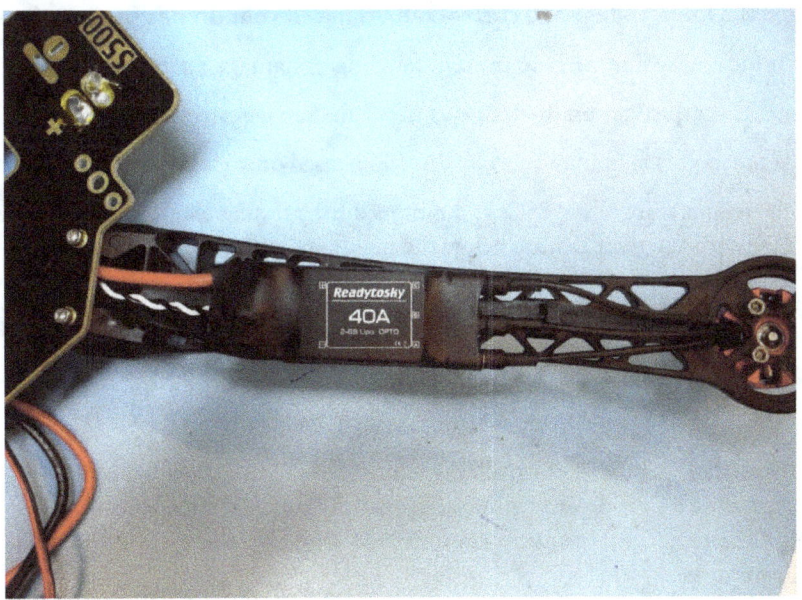

Figure 9.4: ESC mounting

Follow the steps as follows:

1. **Mount the ESCs**: Attach the ESCs to your drone frame, typically on the arms or designated ESC mounting areas. Ensure the drone frame is designed as per the specifications; if it's in a DIY frame, you need to drill or make the mounts at your convenience.

   Secure the ESCs using zip ties, adhesive, or ESC mounting brackets. Ensure they are firmly in place and away from the rotating propellers.

2. **Motor connections**: Connect the motor wires from each motor to their respective ESCs. There should be three wires for each motor (typically labeled as A, B, and C).

   Ensure that the motor's wire colors match the ESC wire colors or follow a consistent wiring scheme for all motors.

3. **Power connections**: Connect the power wires (typically red and black) from the ESCs to the PDB or directly to the battery terminals if your frame allows it.

   Ensure the correct polarity and secure the connections.

4. **Signal connections**: Connect the signal wires (typically white or yellow) from the ESCs to the flight controller. Refer to your flight controller's documentation to determine the appropriate connection points.

   Secure these connections with connectors or soldering and insulate them with heat shrink tubing.

Here are some precautions you should keep in mind:

- Double-check all connections to ensure they are secure and well-insulated
- Ensure that the ESCs' voltage and current ratings match your drone's specifications
- Properly route and secure wires to prevent tangling or damage during flight
- Be cautious of sharp edges on the frame as they can damage wires and ESCs
- Verify that the motors spin in the correct direction based on your chosen motor rotation layout
- Monitor the temperature of the ESCs and motors during testing using an infrared thermometer to ensure they are at a temperature below the manufacturer's recommendations

By following these steps and precautions and conducting thorough testing, you can ensure that your ESCs are correctly assembled, connected, and ready for safe and reliable drone operation.

Installing the flight controller

Mounting and configuring a flight controller on your drone is a crucial step in building a UAV. This section will provide a step-by-step guide on how to mount, configure, test, and verify a flight controller, along with precautions to ensure safe and successful drone operation. The following figure demonstrates this process:

Assembling the components of a drone   179

Figure 9.5: Mounting the flight controller

Follow the steps as follows:

1. **Mount the flight controller**: Locate the designated mounting points on your drone frame where the flight controller will be installed. These points are typically on top of the frame's central plate.

2. **Secure the flight controller**: Attach the flight controller to the mounting points using the provided screws or standoffs. Ensure that the flight controller is mounted securely and is level with the frame. A level flight controller is essential for stable flight.

3. **Wire the connections**: To do this, you must connect the various components to the flight controller:

   - Connect the ESC signal wires to the designated motor output pins on the flight controller
   - Connect the receiver to the flight controller's receiver pins
   - Connect any additional sensors or peripherals (for example, the GPS module) to their respective ports on the flight controller

4. **Power the connections**: Connect the flight controller to the PDB or an appropriate power source on your drone. Ensure proper polarity and secure the connection.

5. **Receiver binding**: For a remote-controlled transmitter and receiver, bind them together according to the manufacturer's instructions. Then, configure the flight controller so that it recognizes the receiver.

Here are some precautions you need to keep in mind:

- Double-check all wiring connections to ensure they are secure and polarized correctly
- Keep the flight controller level during installation for accurate sensor readings
- Ensure the flight controller is well-insulated from vibrations, which can affect sensor accuracy
- Always perform a pre-flight check before each flight to verify the proper operation of the flight controller and all the other components

By following these steps, precautions, and testing procedures, you will be able to mount, configure, and verify your flight controller to ensure that it has been set up for safe and stable drone operation.

Mounting a PDB on a drone

Mounting a PDB on a drone and wiring it is a crucial step in building your UAV. Follow these steps, precautions, and verification methods to ensure a successful assembly:

1. **Identify the mounting points**: Locate the designated mounting points on your drone frame to install the PDB. These points are typically on the central plate. Some frames come with the PDB integrated into the airframe. You might need to put some holes in the PDB.
2. **Position the PDB**: Carefully position the PDB on the mounting points. Ensure it is level with the drone frame for stable flight.
3. **Secure the PDB**: Use the provided mounting hardware (screws and standoffs) to securely attach the PDB to the drone frame. Avoid over-tightening the screws.
4. **Wire the connections (battery)**: Connect the positive (+) and negative (-) wires from your drone's battery to the input terminals on the PDB. Also, ensure proper polarity to prevent short circuits and test this using a multimeter.
5. **ESC connections**: Connect the output terminals of the PDB to the ESCs. These connections distribute power from the PDB to the ESCs, which control the motors.

    Typically, there are labeled outputs on the PDB for each ESC (for example, Motor 1, Motor 2, and so on). Connect the ESCs to their corresponding outputs on the PDB with the correct polarity.
6. **Soldering (if necessary)**: If your PDB requires soldering for the connections, solder the wires carefully to the designated solder pads on the PDB. Use heat shrink tubing to insulate and protect the soldered connections from shorts.
7. **Wire management**: Organize and secure the wires using zip ties or adhesive mounts to prevent tangling and ensure a neat appearance.

8. **Flight controller connection**: Connect the flight controller to the PDB using a power module. Typically, this includes connecting the power (positive and negative), signal wires for ESCs, and other peripherals.

9. **Motor connections**: Connect the motor wires from each motor to their respective ESCs, ensuring the correct order and polarity.

10. **Additional peripheral connections**: If you have additional components, such as cameras, GPS modules, or other sensors, connect them to their respective ports on the flight controller or PDB.

    Here are some precautions you need to take into account:

    - Avoid over-tightening screws and the use of power tools as they could damage the PDB or frame.
    - Organize and secure wires to prevent interference or tangling
    - Ensure that all soldered connections are well-insulated with heat shrink tubing
    - Double-check the polarity of all connections to avoid damage to components
    - Verify that the PDB's voltage and current ratings match the drone's specifications
    - Ensure that exposed wires and connections are insulated to prevent shorts

    Once all the preceding steps are completed, we will need to do some verification.

11. **Visual inspection**: Inspect all connections for loose wires, exposed conductors, or damaged insulation. Then, correct the loose wires and joints and cover them with heat shrinks.

12. **Voltage verification (with a multimeter)**: Use a multimeter to measure the voltage output from the PDB to ensure it matches your battery's voltage when tested.

13. **Continuity check (with a multimeter)**: Verify continuity (low resistance) between the PDB's input and output terminals to ensure there are no open circuits.

By following these steps, precautions, and verification methods, you will be able to successfully mount a PDB on your drone, wire it correctly, and ensure safe and reliable drone operation.

Installing the battery

Assembling and testing the battery on a drone is a crucial step in building and ensuring the safe operation of your UAV. This section provides a step-by-step guide on how to mount and test the battery on the drone, alongside precautions and how to verify the procedure.

Here are some precautions before assembly:

- **Battery compatibility**: Ensure that the battery you plan to use is compatible with your drone in terms of voltage, capacity, and physical size.
- **Battery rating**: Check the battery's voltage (for example, 3S, 4S, 6S, and so on) and capacity (mAh). Ensure it matches the specifications of your drone and its components.

- **Charge state**: If the battery is not pre-charged, charge it using a balance charger. Follow the manufacturer's recommendations for charging rates and storage voltage.

Next, you must do the following for mounting and securing the battery:

1. **Inspect the battery**: Before assembly, visually inspect the battery for any physical damage, punctures, or swelling. Do not use a damaged battery; instead, dispose of it in a disposal bag and drop it off at a recycling point.
2. **Install the connector**: If your battery does not come with a pre-installed connector that matches your drone's power input connector, solder or attach the appropriate connector (for example, XT60, XT90, and so on) to the battery leads. Ensure proper polarity.
3. **Secure the battery**: Use a battery strap or mount to securely fasten the battery to the drone's frame. Make sure the battery is centered and balanced for even weight distribution.
4. **Specify the battery's location**: Position the battery on the drone's frame in a way that maintains the drone's balance and center of gravity. The battery's location can affect flight performance.
5. **Secure the wiring**: Organize and secure the battery wires to prevent them from getting entangled with moving parts or propellers. Use zip ties and ensure proper strain relief.

    Once the preceding steps are completed we need to do testing and verification.

6. **Perform a voltage check (with a multimeter)**: Use a multimeter to check the voltage of the fully charged battery. Ensure it matches the specified voltage (for example, 11.1V for a 3S LiPo and 22.2V nominal voltage for a 6s voltage).
7. **Connect the battery**: Connect the battery to the drone's power input connector while wearing safety goggles. Ensure that you connect the battery with the correct polarity (positive to positive, negative to negative).
8. **Perform a pre-flight voltage check (with a multimeter)**: Before each flight, check the battery's voltage with a multimeter to ensure it is within safe operating limits. LiPo batteries should not be discharged below their specified minimum voltage.

    Here are some precautions you need to consider during and after flight:

9. **Battery temperature**: Monitor the battery's temperature during flight. If it becomes excessively hot, land immediately and inspect the battery.
10. **Discharge limits**: Avoid discharging the battery to a voltage lower than the manufacturer's recommended limit to prevent damage. Typically, it is 3.2V per cell. A key point to note is the allowable charge cycles of the battery by the manufacturer.
11. **Cooldown period**: After flight, allow the battery to cool down before charging or handling it.

By following these assembly steps, precautions, and verification procedures, you can ensure that your drone's battery is assembled correctly, safe to use, and capable of providing reliable power for your UAV's operation.

Configuring the RC transmitter and receiver

Configuring the RC transmitter and receiver with a drone involves mounting, selecting the appropriate protocol, making physical connections, and configuring the system. Always read the manufacturer manual to bind the Tx and Rx with the flight controller.

The steps for mounting the RC transmitter are as follows:

1. **Strap or harness**: Attach the neck strap or harness to the transmitter if it's included. This allows you to wear the transmitter comfortably and have your hands free for control.
2. **Grip**: Hold the transmitter comfortably with a firm grip. Ensure your fingers have easy access to the control sticks and switches.
3. **Antenna position**: Keep the transmitter's antenna pointed upward and avoid obstructing it with your body or clothing to maintain a clear signal path to the drone.

The steps for mounting the RC receiver are as follows:

1. **Secure location**: Mount the RC receiver securely inside the drone's frame or designated receiver compartment. Ensure it's protected from vibration, moisture, and physical damage.
2. **Antenna placement**: Extend the receiver's antennas outside the drone's frame. Place them as far apart as possible for diversity and avoid having them close to metal components or carbon fiber, which can cause signal interference.
3. **Antenna orientation**: Orient the receiver antennas perpendicular to each other (typically at a 90-degree angle) for optimal signal reception.
4. **Secure fastening**: Use zip ties, Velcro straps, or foam padding to secure the receiver in place. Ensure it cannot come loose during flight.
5. **Antenna routing**: Route the receiver antennas in a neat and organized manner, away from moving parts, propellers, or anything that might damage them.

By following these steps for mounting, protocol selection, connecting, and configuring, you can establish a reliable and responsive control link between your RC transmitter and receiver and your drone. Always consult the user manuals and documentation provided by the transmitter and receiver manufacturers for specific setup and configuration details.

Telemetry connection

Long-range telemetry systems are essential for maintaining communication with your drone over extended distances. This section explains mounting, selecting a protocol, connecting, and configuring when using a long-range telemetry system.

Mounting long-range telemetry components

You'll need two telemetry modules – one for your drone and one for your ground station. These modules are typically designed to be compact and lightweight for ease of mounting.

Each telemetry module requires an antenna. Mount the antennas in locations that provide a clear line of sight to each other. Higher placement and less obstruction offer better signal quality.

You must securely mount the telemetry modules and antennas on your drone and ground station using suitable mounting hardware, such as brackets, clamps, or adhesive mounts.

Selecting a protocol

Long-range telemetry systems often use protocols such as **Long Range Systems (LRS)**, MAVLink, or proprietary protocols specific to the manufacturer. Ensure compatibility between your modules and choose a protocol that suits your requirements. Generally, within the scope of this chapter, we are considering the serial protocol with RFD900 telemetry modules, which we studied earlier.

Connection and configuration

Here is how to connect and configure a long-range telemetry system with your drone, specifically using an LRS system as an example:

1. **Hardware connection**: Connect one telemetry module to your drone's flight controller. This typically involves connecting the telemetry module's TX and RX wires to a free UART port on the flight controller. Ensure that the voltage levels are compatible (usually 3.3V or 5V).

   Then, connect the other telemetry module to your ground station's computer or telemetry device via USB or a compatible serial port.

2. **Configuring the ground station**: Install the ground station software or telemetry configuration software provided by the manufacturer on your computer. For RFD tools, there is software named *RFD tools* that you can use to configure and bind the telemetry radios.

   Then, connect the ground station telemetry module to your computer. Once you've done this, you must configure the telemetry module's settings using the manufacturer's software. This may include selecting the appropriate protocol, configuring the air and ground unit IDs, and adjusting transmission power settings.

   Now we will configure the drone (flight controller)

3. Power on your drone and connect it to the ground station via USB or telemetry.

4. Open the configuration software for your drone's flight controller (for example, Betaflight, ArduPilot, and so on).

5. Configure the telemetry protocol and port so that they match your telemetry module's settings. Ensure that the baud rate and other settings match between the flight controller and telemetry module.

6. Enable telemetry or MAVLink data transmission on the flight controller. This step is not necessarily required since some flight controllers are pre-configured. This typically involves enabling a telemetry port and setting its parameters.
7. Write the configuration changes to the flight controller.

   We will then do testing and verification.
8. Power up your drone and ground station.
9. Check that telemetry data is being transmitted from the drone to the ground station and vice versa.
10. Monitor telemetry data on your ground station software, such as GPS position, altitude, battery voltage, and other relevant information.

By following these steps and using the appropriate long-range telemetry components and protocols, you can successfully configure a reliable long-range telemetry communication link with your drone. Remember to consult the user manuals and documentation provided by the telemetry system manufacturer for specific setup and configuration details.

Setting up and configuring avionics

The avionics configuration of a drone refers to the suite of electronic systems and components that are responsible for its navigation, control, and communication. This critical aspect of drone design typically includes GPS receivers for accurate positioning, flight controllers for stabilization and control, telemetry systems for data transmission, sensors such as accelerometers and gyroscopes for flight data acquisition, and often advanced features such as obstacle avoidance sensors and cameras for autonomous navigation and data capture. Avionics play a vital role in ensuring safe and precise flight operations, making them an integral part of modern drone technology.

Firmware flashing

Flashing firmware to a flight controller using **Aero Ground Control Station** (**AeroGCS**) for a quadcopter is a crucial process to ensure that your flight controller has the latest software and features. Here's a general guide on how to flash firmware using AeroGCS:

1. **Prepare your equipment**: Ensure that your quadcopter is powered off and disconnected from the AeroGCS-equipped device (for example, your laptop or tablet).
2. **Launch AeroGCS**: Open AeroGCS on your device.
3. **Connect the flight controller**: Connect your flight controller to your device using the appropriate communication method (USB, telemetry radio, and so on). Ensure that the connection is stable.
4. **Access the firmware update menu**: In AeroGCS, navigate to the firmware update or flash firmware menu. The location of this menu may vary, depending on the AeroGCS version you're using.

5. **Select the firmware**: Select the firmware file for your flight controller. Select the file to upload it to AeroGCS.

6. **Flash firmware**: Follow the onscreen instructions to initiate the firmware flashing process. Typically, this involves clicking on a **Flash** or **Update** button.

7. **Verify the firmware's version**: Once the firmware update is successful, verify that the flight controller now has the updated firmware version. This information is usually displayed within AeroGCS.

8. **Test and verify**: After the firmware update, conduct a ground test to ensure that the flight controller is operating correctly. Verify that all sensors and functions are working as expected.

Configuring the flight controller

Configuring a flight controller for a quadcopter using AeroGCS typically involves several key parameters and settings to ensure safe and precise flight operations. Here's a list of common configurations that are usually required and a general guide on how to do them using AeroGCS:

Motor layout and propeller configuration:

Purpose: Specify the motor layout and propeller type so that they match your quadcopter's physical configuration

How to do it:

1. Connect your quadcopter to AeroGCS.
2. Access the motor layout and propeller configuration settings.
3. Select the appropriate motor layout (for example, X-configuration, H-configuration, and so on) and enter details about your quadcopter's propellers, including diameter and pitch.

Flight mode configuration:

Purpose: Define different flight modes (for example, stabilize, altitude hold, GPS position hold, and **return-to-home** (**RTH**)) and assign them to transmitter switches or control channels.

How to do it:

1. Access the flight modes configuration menu.
2. Assign flight modes to specific transmitter switches or control channels.
3. Configure the parameters for each flight mode, if necessary.

RC transmitter calibration:

Purpose: Ensure that your quadcopter responds accurately to transmitter inputs

How to do it:

1. Connect your drone to AeroGCS.
2. Access the RC transmitter calibration menu.
3. Calibrate the transmitter endpoints and verify that all control channels operate correctly.

Safety and failsafe configuration:

Purpose: Establish failsafe mechanisms to ensure safe operation in case of signal loss or critical errors.

How to do it:

1. Access the safety and failsafe configuration menu.
2. Define actions to be taken during signal loss, low battery, or other emergencies (for example, RHT or auto-land).

Sensor calibration:

Purpose: Calibrate onboard sensors, such as the accelerometer, gyroscope, and compass, to ensure accurate flight data.

How to do it:

1. Access the sensor calibration menu.
2. Follow the onscreen instructions to calibrate each sensor, usually by placing the quadcopter in specific orientations.

Battery configuration:

Purpose: Configure your battery settings to monitor voltage and current accurately.

How to do it:

1. Access the battery configuration menu.
2. Set the battery voltage and current sensor parameters.
3. Specify low voltage warnings and critical voltage levels.

ESC calibration:

Purpose: Ensure that all motors spin at the same speed when given the same throttle input.

How to do it:

1. Access the ESC calibration menu.
2. Follow the provided instructions, which may involve arming the motors and performing throttle range calibration.

These are the fundamental configurations that are typically required for a quadcopter's flight controller using AeroGCS. Please note that specific options and procedures may vary based on your flight controller's firmware, AeroGCS version, and the quadcopter's hardware. Always consult the user manuals and documentation provided with your equipment for detailed and accurate configuration procedures tailored to your specific setup.

Understanding calibration and failsafes

Sensor calibrations are required and are a crucial step in pre-flight preparation. This enables the various sensors, such as the IMU, barometer, and magnetometer, to fetch the initial values for the heading, height, and other data points that are crucial for flight.

Calibrating sensors

Calibrating various sensors and components in a drone is essential to ensure accurate and stable flight performance. The specific calibrations required may vary based on your drone's hardware and flight controller, but here are some common calibrations and how to perform them using AeroGCS:

Accelerometer calibration:

Purpose: To ensure the accurate measurement of acceleration, which is essential for stabilization and attitude control.

How to do it:

1. Connect your drone to AeroGCS.
2. Navigate to the accelerometer calibration menu in AeroGCS.
3. Follow the onscreen instructions, which typically involve placing the drone at various orientations and allowing it to settle while AeroGCS records sensor data.

Gyroscope calibration:

Purpose: To ensure the accurate measurement of angular velocity, which is crucial for controlling the drone's orientation.

How to do it:

1. Connect your drone to AeroGCS.
2. Access the gyroscope calibration menu in AeroGCS.
3. Follow the provided instructions, which often involve keeping the drone stationary while AeroGCS collects data.

Compass calibration:

Purpose: To calibrate the magnetometer or compass sensor to provide accurate heading information.

How to do it:

1. Choose the compass calibration option in AeroGCS.
2. Typically, you'll be prompted to rotate the drone along multiple axes while following onscreen instructions.
3. Avoid magnetic interference during calibration, such as metal objects or electronic devices.

RC transmitter calibration:

Purpose: To ensure that your drone responds accurately to transmitter inputs.

How to do it:

1. Connect your transmitter to AeroGCS.
2. Access the RC transmitter calibration menu in AeroGCS.
3. Follow the instructions to calibrate your transmitter endpoints and ensure that all control channels operate correctly.

GPS configuration:

Purpose: To configure and calibrate the GPS for accurate position and altitude information (if applicable).

How to do it:

1. Enable and configure the GPS in AeroGCS.
2. Set up a home location and configure GPS-related flight modes, such as position hold and RTH.

ESC calibration:

Purpose: To ensure that all motors spin at the same speed when given the same throttle input.

How to do it:

1. Access the ESC calibration menu in AeroGCS.
2. Follow the instructions, which typically involve arming the motors and then performing throttle range calibration.

Battery calibration:

Purpose: To calibrate the flight controller's battery voltage monitoring for accurate battery level readings.

How to do it:

1. Access the battery calibration menu in AeroGCS.
2. Follow the provided instructions, which may involve fully charging and discharging the battery while AeroGCS records voltage values.

Level calibration:

Purpose: To ensure that the drone remains level when flying in stable modes.

How to do it:

1. Access the level calibration menu in AeroGCS.
2. Place the drone on a flat, level surface and follow the onscreen instructions to calibrate the horizon reference.

It's important to follow the specific instructions provided by AeroGCS for each calibration procedure as they may vary based on your flight controller's firmware and AeroGCS version. Additionally, perform these calibrations in a controlled and interference-free environment for the most accurate results. Always refer to your drone's user manual and AeroGCS documentation for detailed calibration procedures tailored to your equipment.

Failsafe setup and configuration

Failsafes are critical safety mechanisms in a quadcopter that automatically trigger predefined actions to ensure safe flight in the event of unexpected events or system failures. Setting up failsafes in a quadcopter using AeroGCS is essential to prevent accidents and minimize damage. Here are some common failsafes, how to set them up in AeroGCS, and how the drone behaves when they are triggered:

RTH failsafe:

Purpose: In the event of signal loss or another critical issue, the quadcopter will automatically return to its home position, usually where it took off.

Behavior when triggered: The drone will ascend to the specified RTH altitude. It will then fly back to the home point using GPS. Once it reaches the home point, it may hover or descend to land, depending on your configuration.

How to set it up in AeroGCS:

1. Access the failsafe configuration menu in AeroGCS.
2. Enable the RTH failsafe option.

3. Specify the RTH altitude and climb/descent rates.
4. Set the RTH mode to either **Smart RTH** (uses GPS) or **Simple RTH** (uses the original takeoff direction).

Land failsafe:

Purpose: If the quadcopter encounters a failsafe condition, it will land at its current location.

Behavior when triggered: The drone will initiate a controlled descent and land at its current location.

How to set it up in AeroGCS:

1. In the failsafe configuration menu, enable the **Land failsafe** option.
2. Set the altitude and descent rate for landing.

Motor kill failsafe:

Purpose: In the event of a critical issue, the quadcopter will cut power to the motors, causing an immediate shutdown.

Behavior when triggered: The drone will immediately cut power to the motors, causing it to drop from the sky.

How to set it up in AeroGCS:

1. Access the failsafe configuration menu.
2. Enable the **Motor Kill failsafe** option.
   - Set the time delay for the motor kill action (if applicable).

Low battery failsafe:

Purpose: When the battery voltage drops below a specified threshold, the quadcopter will trigger a failsafe action to prevent a crash.

Behavior when triggered: The drone will execute the specified action, which could be RTH, landing, or hovering in place.

How to set it up in AeroGCS:

1. In the failsafe configuration menu, enable the **Low Battery failsafe** option.
2. Specify the voltage level at which the failsafe should be triggered.
3. Set the action to be taken, such as RTH, land, or hover.

GPS glitch failsafe:

Purpose: In case of GPS signal anomalies or glitches, the quadcopter can be programmed to take specific actions to maintain stability.

Behavior when triggered: The drone will follow the configured behavior, which may include maintaining its current altitude and position or switching to an alternative flight mode.

How to set it up in AeroGCS:

1. Access the failsafe configuration menu.
2. Enable the **GPS Glitch failsafe** option.
3. Configure the desired behavior, such as maintaining altitude and position or switching to an alternative mode.

RC signal loss failsafe:

Purpose: If the RC transmitter signal is lost or weak, the quadcopter will initiate a failsafe action to regain control.

Behavior when triggered: The drone will execute the configured failsafe action, which might involve RTH, landing, or hovering in place, depending on your settings.

How to set it up in AeroGCS:

1. In the failsafe configuration menu, enable the **RC Signal Loss failsafe** option.
2. Specify the timeout duration for signal loss.

The specific behavior of the drone when a failsafe is triggered can vary based on your configuration. It's essential to configure failsafes carefully so that they align with your safety and operational requirements. Always test failsafes in a controlled environment to ensure they function as expected and to understand how your quadcopter responds in various scenarios.

Setting up a maiden flight

With everything set up, we can start setting up the drone's maiden flight. Note that you must be extra careful during the first flight and observe the drone's behavior. Make sure you follow the safety guidelines for yourself and the people around you. It's always recommended to do the maiden flight in an open area so that fewer people are around. Make sure that the drone is away from potential buildings, structures, and populations. The safety of the public is important and we should be responsible pilots.

Setting up flight modes, a **return-to-land** (**RTL**) switch, and joystick controls on your transmitter for flying a drone requires proper transmitter configuration. Here's a step-by-step guide on how to set up these functions:

1. **Identify the transmitter controls**: Familiarize yourself with the control switches and channels on your transmitter. Typically, you'll have switches, buttons, and control sticks (joysticks) available.
2. **Access the transmitter's configuration menu**: Depending on your transmitter model, access the transmitter's configuration or setup menu. Consult your transmitter's manual for specific instructions on accessing this menu.
3. **Assign channels to the different flight modes**: Flight modes determine how the drone behaves during flight. Assign specific switches or buttons on your transmitter to toggle between different flight modes (for example, stabilize mode, altitude hold, GPS position hold, and RTL).
4. **Configure flight modes in AeroGCS**: Open AeroGCS on your device. Access the flight modes configuration menu in AeroGCS.

    Define the various flight modes you want to use and assign them to specific transmitter switches or channels. Here are some examples:

    - **Stabilize mode**: Assign a switch/button for manual flight
    - **Altitude hold mode**: Assign another switch/button to maintain a consistent altitude
    - **GPS position hold mode**: Assign a switch/button for GPS-assisted position hold
    - **RTL mode**: Assign a dedicated switch/button for RTL

5. **Set up the RTL switch**: Designate a switch on your transmitter as the RTL switch. This switch will trigger the RTL function. Then, configure the RTL switch in AeroGCS to activate RTL mode when the designated switch is flipped or pressed.
6. **Configure joystick controls**: If you want to use the control sticks (joysticks) on your transmitter to manually control the drone, you can assign the appropriate control channels. Assign the following functions to the respective control sticks:

    - **Throttle (altitude control)**: Assign the throttle control to the left joystick's up/down movement
    - **Yaw (rotation control)**: Assign yaw control to the left joystick's left/right movement
    - **Pitch (forward/backward tilt)**: Assign pitch control to the right joystick's up/down movement
    - **Roll (sideways tilt)**: Assign roll control to the right joystick's left/right movement

7. **Save the transmitter's configuration**: After assigning flight modes, the RTL switch, and joystick controls, save the transmitter's configuration in both the transmitter itself and AeroGCS. This ensures that the settings are retained for future flights.

8. **Conduct pre-flight checks**: Before each flight, perform a pre-flight check to ensure that all controls respond as intended and that the drone behaves correctly in response to your transmitter's inputs.

The precise steps and options for setting up these functions may vary depending on your specific transmitter and AeroGCS setup. Always refer to the user manuals for your transmitter and AeroGCS for detailed and accurate instructions tailored to your equipment. Additionally, follow safety guidelines and local regulations when flying your drone.

Performing a maiden flight

Performing the first manual flight of your drone using AeroGCS and a transmitter involves several steps to ensure a safe and controlled flight experience. Here's a general guide on how to do your first flight manually:

1. **Pre-flight preparation**:

    - **Check local regulations**: Ensure that you are aware of and compliant with local regulations regarding drone flight. Obtain any necessary permits or authorizations if required.

    - **Choose a suitable location**: Select a wide-open, open-air space free from obstacles, people, and other potential hazards. Avoid flying near airports or in restricted airspace.

    - **Inspect your drone**: Conduct a thorough pre-flight inspection of your drone, including checking for loose screws, damaged propellers, secure battery connections, and overall airworthiness.

    - **Check the battery's status**: Ensure the flight battery is fully charged and properly connected. Confirm that all battery indicators are within the safe operating range.

    - **Calibrate the sensors**: Calibrate onboard sensors such as the accelerometer and compass according to the manufacturer's instructions and as needed for your specific drone.

2. **Power on your drone**: Power on your drone and wait for it to initialize. Make sure it establishes a GPS lock if you plan to use GPS-assisted flight modes.

3. **Connect your transmitter to AeroGCS**: Turn on your transmitter and ensure it is properly bound to your drone's receiver. Then, connect your computer or device running AeroGCS to your drone. Establish a stable telemetry link between AeroGCS and your drone if you plan to monitor flight data and telemetry.

4. **Perform a transmitter controls check**: Test the controls by moving the control sticks (throttle, yaw, pitch, and roll) and toggling the switches on your transmitter. Ensure that all controls respond correctly and smoothly.

5. **Arm the drone**: Depending on your drone and AeroGCS configuration, you may need to arm the drone before takeoff. Follow the instructions provided to arm your drone safely.

6. **Take off manually**: Carefully apply the throttle to lift the drone off the ground. Ensure that you maintain control and stability during takeoff.
7. **Practice basic flight maneuvers**: Start with simple maneuvers to get a feel for the controls. Begin with gentle pitch and roll movements while maintaining altitude. Then, gradually increase the complexity of your flight maneuvers as you become more comfortable with the controls. Practice yaw (rotation) and coordinated turns.
8. **Monitor flight data**: Continuously monitor the telemetry data displayed in AeroGCS to ensure that all systems are functioning correctly and that you are within safe operating parameters.
9. **Fly at a safe altitude**: Maintain a safe and controlled altitude during your first flights. Avoid flying too high or too close to the ground to minimize the risk of accidents.
10. **Land the drone**: When you are ready to land, initiate a gentle descent while reducing the throttle. Land the drone softly and safely.
11. **Post-flight inspection**: After landing, conduct a post-flight inspection of your drone to ensure there is no damage or issues that need attention.

Flying a drone manually requires practice, especially if you are new to piloting drones. Start with basic maneuvers and gradually progress to more advanced techniques as you gain confidence and experience. Always prioritize safety and abide by local regulations when flying your drone.

Summary

In this chapter, you learned about drone assembly and gained some practical experience in doing this so that you have a deeper understanding of drone technology. You gained hands-on insights into the intricate components and systems that make UAVs function effectively. This knowledge can be invaluable for drone enthusiasts, hobbyists, and professionals alike. Apart from the educational aspect, mastering drone assembly enables users to develop their UAVs to specific needs and preferences, potentially leading to enhanced performance and versatility.

In the next chapter, we'll consider another aspect of drone assembly and how it can be more tailor-made.

10
Flight Log Analysis and PIDs

In the previous chapter, we studied setting up a drone from end to end, configuring it, and flying it. This process can be quite tricky initially and require further research and practice, but once we are used to it, it can be a fun everyday activity. In this chapter, we will study the analysis of the performance of sensors and actuators of drones during a flight, along with **Proportional-Integral-Derivative (PID)** gains, and make the required changes for a better, smoother flight. Analyzing flight logs and understanding PID controllers are critical skills for beginners in the world of drone flight. After your first drone flight, log analysis and PID tuning become essential tools for improving your piloting skills and enhancing your drone's performance. PID controllers are at the core of **flight control systems (FCSs)**, responsible for maintaining the drone's stability and responsiveness. Understanding PID principles and how to fine-tune these controllers enables beginners to optimize flight performance, reduce oscillations, and ensure precise control of their drone.

Moving ahead with this chapter, we are going to cover the following main topics, driving through them one by one sequentially to understand their connectivity in flight performance:

- Introduction to flight logs and their applications
- Working with **Ground Control Station (GCS)** software on logs and their graphs
- Understanding PID controllers and their uses

Technical requirements

Log analysis tools are different for flight controllers using open source firmware. It typically involves common software and hardware tools, regardless of the specific firmware. The skills required for log analysis are also similar across various open source flight control platforms. Here's an overview of the requirements:

- **Log analysis software**: Many open source firmware solutions provide log analysis tools as part of their ecosystem, such as the following:

    - **PX4**: Flight Review (`https://review.px4.io/`), pyulog (`https://github.com/PX4/pyulog`).
    - **ArduPilot**: Mission Planner, APM Planner 2.0, MAVProxy, MAVExplorer, and others.
    - **GCS**: GCS software, such as Mission Planner, **QGroundControl** (**QGC**), APM Planner, or AeroGCS, is necessary to download, view, and analyze flight logs.

- **Graphing and data analysis software**: Additional tools for in-depth data analysis, visualization, and graphing can be beneficial. Tools such as Microsoft Excel, MATLAB, or Python with libraries such as Matplotlib or pandas are useful.

- **Data interpretation**: The ability to interpret graphs, charts, and numerical data is crucial. Understanding how to correlate data from various sensors and flight parameters is essential.

- **Flight dynamics knowledge**: A basic understanding of drone flight dynamics is necessary, including how sensors, motors, and control surfaces impact flight behavior.

- **PID controller understanding**: Knowledge of PID controllers and how they influence drone response is required, especially if you want to analyze and fine-tune PID settings.

Introduction to flight logs and their applications

Drone flight logs are digital records that capture maximum information about a drone's flights. They contain every detail, such as flight duration, GPS coordinates, altitude, battery status, and sensor readings. These logs are important as they provide a detailed history of a drone's activities and help engineers diagnose issues, optimize flight parameters, and ensure safe operations. Being drone engineers, we can use these logs for data analysis, performance assessment, and predictive maintenance, which all contribute to enhanced drone reliability and efficiency. In essence, drone flight logs are indispensable tools for understanding, improving, and maintaining drone systems.

Drone logs are essential in solving and analyzing multiple factors that impact the performance of the drone and the performance of the system avionics assisting the drone to fly. Some common issues that can be identified through the analysis of drone logs include the following:

- **Flight errors**: Logs can reveal flight errors, such as erratic movements, sudden drops in altitude, or deviations from the planned flight path, which could indicate a problem with the drone's flight controller or sensors.

- **Battery problems**: Logs track battery voltage, current, and capacity, allowing operators to identify battery issues such as poor performance, excessive discharges, or an aging battery that may need replacement.
- **GPS issues**: Logs record GPS coordinates so that discrepancies or disruptions in the GPS signal can be identified. This is crucial for maintaining accurate position data during flights.
- **Temperature and environment**: Logs can show if the drone was exposed to extreme temperatures or environmental conditions that might have affected its performance or sensors.
- **Motor or propeller failures**: Anomalies in motor speed or abnormal changes in thrust can indicate motor or propeller problems that may need attention.
- **Sensor failures**: Drone logs can reveal sensor issues, such as gyroscopic or accelerometer anomalies, which are critical for stable flight.
- **Communication problems**: If there are disruptions in communication between the drone and the remote controller, logs can help pinpoint the source of the issue, whether it's interference or radio signal problems.
- **Wind and weather effects**: By analyzing drone logs, operators can assess how wind and weather conditions impact flight performance and stability, aiding in safer operations.
- **Data for maintenance**: Regularly examining drone logs can assist in predictive maintenance by identifying wear and tear on components such as motors, batteries, and propellers before they fail.
- **Flight path analysis**: Drone logs can be used to review the flight path to ensure that the drone followed the planned route and adhered to airspace restrictions, preventing violations and unauthorized flights.

Flight logs are stored within a drone's onboard memory or storage medium, such as an SD card. These logs are saved in digital formats, most commonly as plain text or **comma-separated values** (**CSV**) files. CSV files, as the name suggests, are files containing plain text files that store data separated by commas. These files help to organize the data, which helps in smooth data readings. To access and read these logs, we connect the drone to a computer via a USB cable or remove the storage medium and insert it into a card reader. Once the logs are physically accessible, compatible software or tools can be employed for analysis. This software may be drone-specific or third-party applications designed for log analysis, as we witnessed earlier. As we have studied, the logs provide a wealth of information, including flight duration, GPS coordinates, altitude, battery status, and sensor readings, which can be interpreted to understand the drone's performance and behavior during flights. Properly managing and analyzing flight logs is important for improving flight efficiency, ensuring regulatory compliance, diagnosing issues, and maintaining the overall health of the drone.

Types of logs stored

There are various types of logs recorded with different aspects of their operations. Here's a brief explanation of the main types of drone logs, their uses, and how to access them:

- **Flight logs**:

    Use: Flight logs are comprehensive records of each flight, including details such as altitude, GPS coordinates, speed, battery status, motor performance, and sensor readings. They offer a detailed overview of the drone's behavior during a specific flight.

    Access: Flight logs are typically accessed through the flight controller, where we can review and export them for analysis.

- **Telemetry logs**:

    Use: Telemetry logs focus on real-time data, providing information such as battery voltage, current draw, GPS signal strength, and sensor readings. They help monitor the drone's performance during flight. These are reflected on the GCS for the pilots to view.

    Access: Telemetry data is often accessible in real time through the drone's remote controller or ground station software, which can be accessed later.

- **Error logs**:

    Use: Error logs are generated when the drone encounters issues or errors during a flight. They detail problems such as motor failures, sensor malfunctions, or communication issues, assisting in issue diagnosis.

    Access: Error logs can usually be accessed through the drone's companion app or flight controller software, specifically in sections related to error reporting and diagnostics.

- **Mission logs**:

    Use: Mission logs are specific to programmed tasks or missions. They capture data related to waypoints, actions, and mission-specific parameters, allowing for a review of the execution of predefined tasks.

    Access: Mission logs are often available through mission planning and execution software used with the drone.

- **Control logs**:

    Use: Control logs document the drone's control inputs, including commands from the remote controller or ground station. They help in understanding how the drone was operated during a flight.

Access: Control logs can be accessed through the remote controller or specific control software, where you can review pilot inputs and commands.

We have seen different types of logs and what data they store. In the following sections, we will look at a method to read those logs and extract meaningful data from them.

Working with GCS software on logs and their graphs

Flight logs are generated during the flight and stored in the internal storage and memory cards of the drone. In the following section, we will explore how to upload logs to a specific tool and access them using the QGC software and Flight Review.

Steps to download logs from the flight controller using QGC as the GCS

The latest build of QGC can be downloaded and installed from their official website. It's open source GCS software useful for the setting up and configuration of drones for manual and mission flights. The QGC software is compatible with Linux, Windows, and Mac on a PC and is also compatible with iOS and Android devices. It supports flight stacks such as ArduPilot, PX4, or any other stack that communicates through MAVLink. The detailed steps to download and read the logs are as follows:

1. **Connect your drone**: Connect your drone to your computer using a USB cable or a telemetry module, ensuring it's powered on and connected to the QGC software.
2. **Launch QGC**: Open the QGC software on your computer. If you haven't installed it, you can download it from the official QGC website.
3. **Establish a connection**: In QGC, establish a connection to your drone. This can often be done through the **Connect** or **Vehicle** menu, depending on your QGC version.
4. **Access the log section**: Navigate to the **Log** or **Data Analysis** section within QGC. This section may also be labeled **Telemetry Logs**, **Download Logs**, or something similar, depending on your QGC version. The following screenshot of the log download window of QGC shows data for the logs stored:

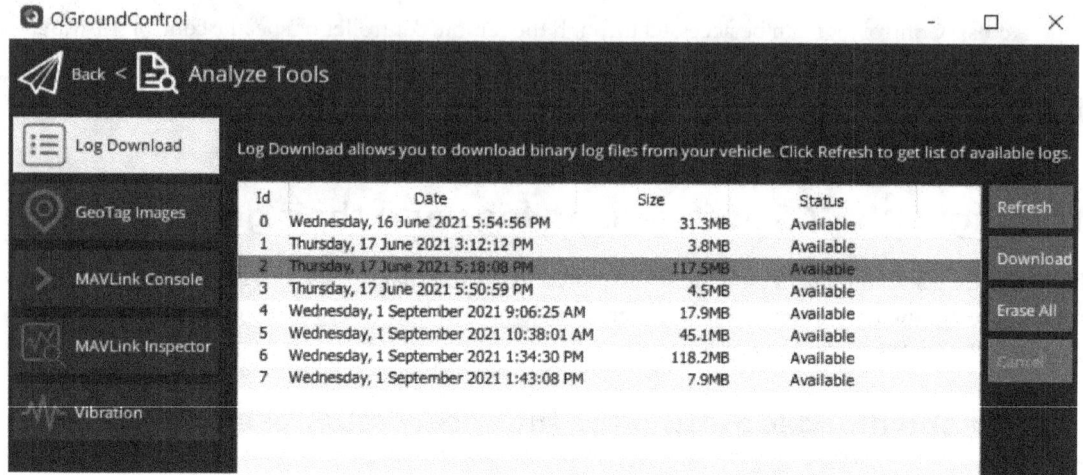

Figure 10.1 – Log download section of QGC

5. **Select the logs to download**: In the **Log** section, you will see a list of available logs. Select the logs you want to download. These logs are typically organized by date and time.

6. **Initiate download**: Once you've selected the desired logs, look for an option that allows you to download or retrieve the selected logs. This option may be labeled **Download Logs**, **Save Logs**, or something similar.

7. **Choose a download location**: QGC will prompt you to specify the location on your computer where you want to save the downloaded logs. Select an appropriate folder or directory.

8. **Access the downloaded logs**: Once the download is complete, you can access the downloaded logs from the location you specified in *step 7*. The logs are typically saved as .ulg files (UAV log format), which can be opened and analyzed using log analysis tools.

9. **Review and analyze**: To review and analyze the logs, you can use log analysis software, including dedicated log analysis tools or web-based platforms such as Flight Review, as previously mentioned in the context of QGC.

Analyzing download logs

Since we are using QGC for downloading logs, the logs are saved in the .ulg file format, which is best read and analyzed by the Flight Review tool. There are many other tools available, such as PlotJuggler, which can be used for in-depth estimation. Analyzing a log using Flight Review, a web-based log analysis tool commonly associated with the PX4 flight stack, involves reviewing various data and graphs to assess your drone's performance during a flight. Here's a brief overview of how to analyze logs using Flight Review and the different graphs it provides:

1. **Access Flight Review**: Open a web browser and go to the Flight Review website (https://logs.px4.io/).

2. **Upload and analyze log files**: Click the **Upload** or **Analyze** button on the Flight Review website:

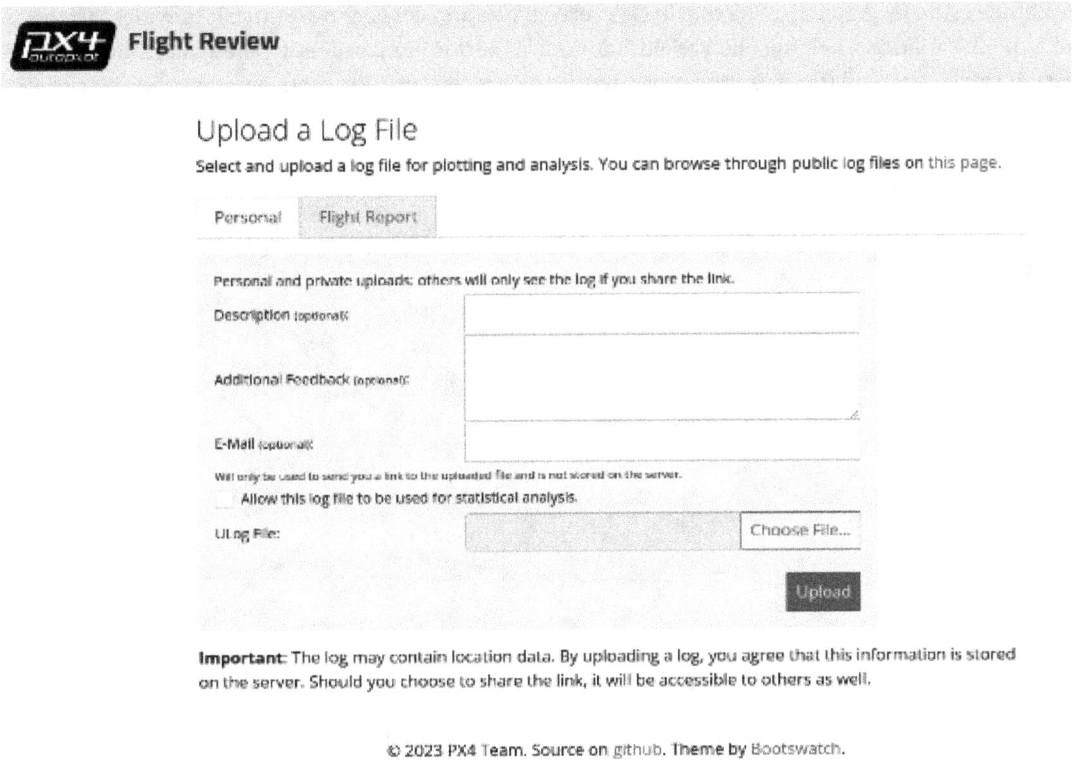

Figure 10.2 – Flight Review home page

Select the log files you want to analyze from your computer and upload them.

After you upload the log file, it is analyzed by the tool, and each data point is presented in the form of a value-time graph that can be read and interpreted.

Let's see next how to read and interpret graphs.

Interpreting graphs

Different flight modes (for example, stabilize, altitude hold, loiter, and return to home) may be represented by distinct colors in the graph. This helps you visually identify which flight mode the drone was in at any given time.

Altitude estimate

An altitude estimate graph displays the drone's vertical position changes over time. It provides insights into altitude variations, helping analyze altitude control performance, identify anomalies, and ensure mission safety, particularly in applications requiring precise altitude management, such as aerial mapping and surveying:

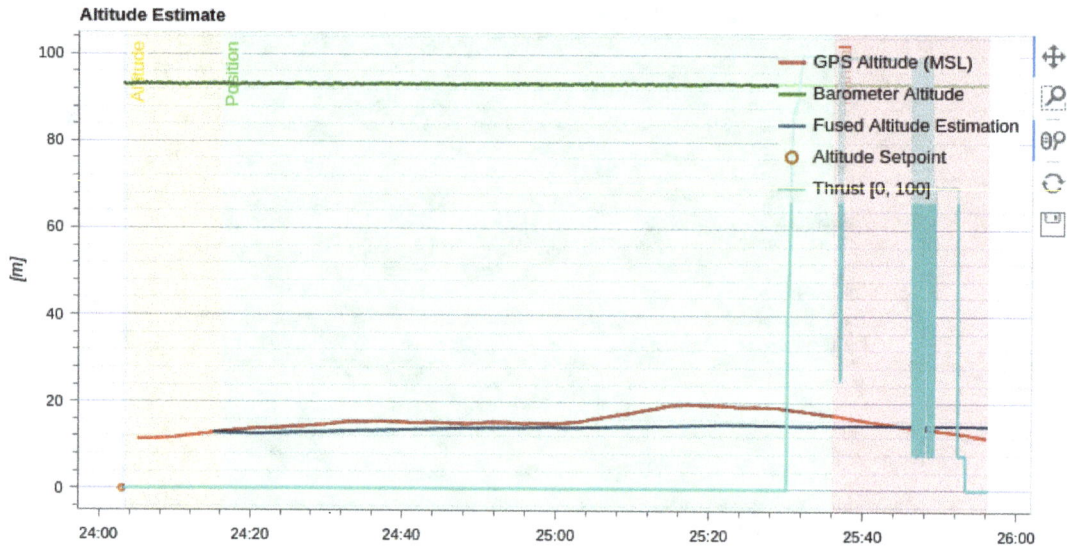

Figure 10.3 – Altitude estimate graph

Attitude (roll, pitch, yaw, and angular rates) estimate

Roll, pitch, yaw, and angular rate graphs display the drone's attitude and angular movement. They help assess how the drone is oriented and how it's maneuvering in terms of roll, pitch, and yaw angles. An angular rate graph shows the rate of change in these angles. These graphs provide critical information about the drone's stability, responsiveness, and adherence to control inputs, enabling performance analysis and troubleshooting. The following is an example of a flight attitude estimate graph:

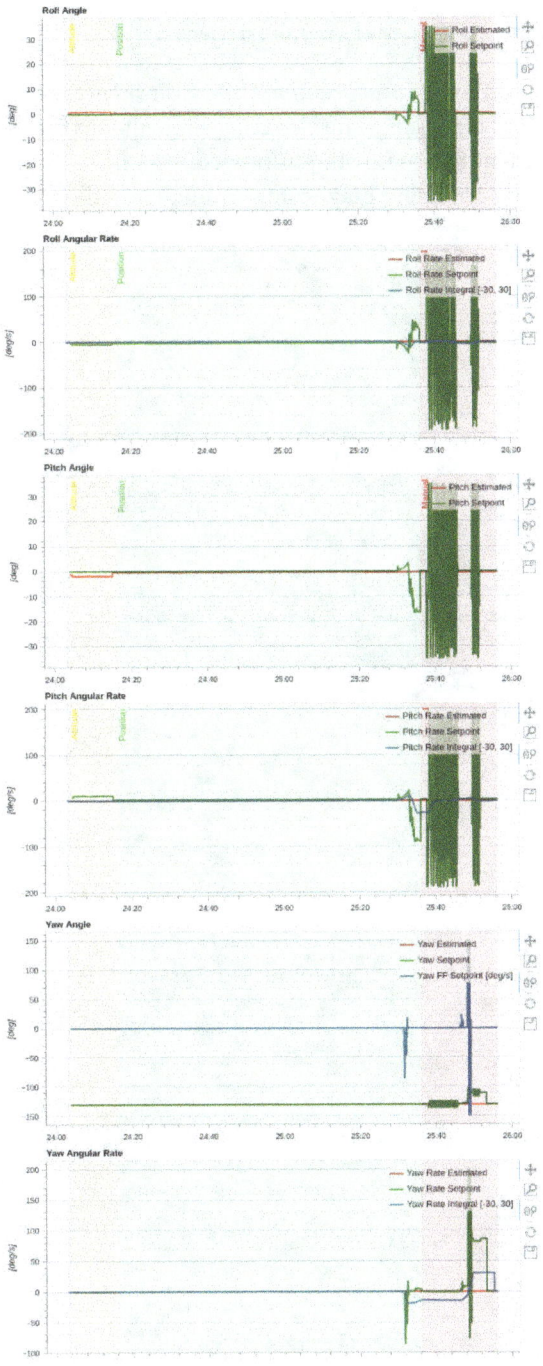

Figure 10.4 – Flight attitude estimate graph

Local position estimate (X, Y, Z) estimate graph

Local Position X, *Local Position Y*, and *Local Position Z* graphs display the drone's position in its local coordinate system. These graphs are essential for detecting anomalies in the drone's lateral and vertical movement, which can help identify issues with position control, navigation errors, or disturbances. Analyzing these graphs can be critical in diagnosing anomalies and ensuring the drone's safe and precise operation:

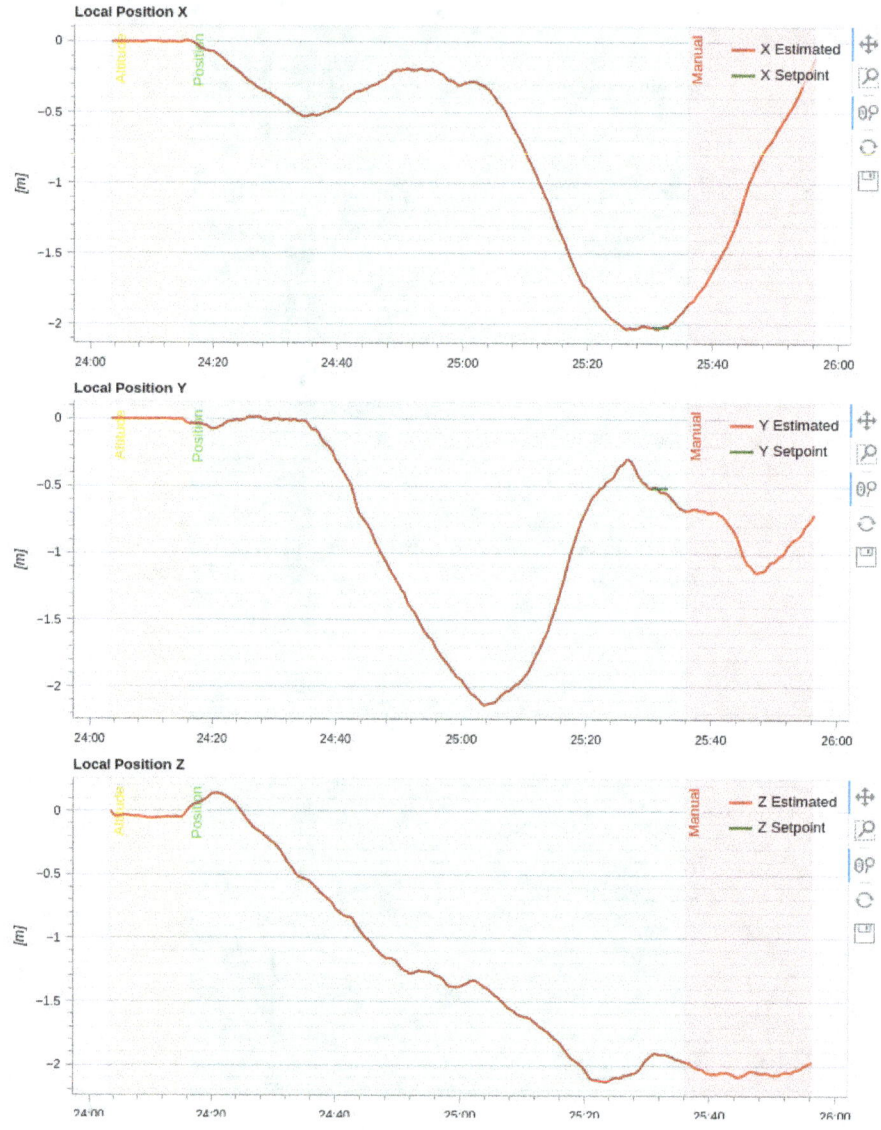

Figure 10.5 – Local position estimate graph

Actuator output graph

An actuator output graph provides insight into how the drone's control surfaces and motors are responding to commands. This data is crucial for detecting anomalies such as unexpected motor behavior, which can indicate issues with flight stability, control algorithms, or mechanical problems. Analyzing this graph helps in diagnosing anomalies and ensuring the drone's safe and stable operation:

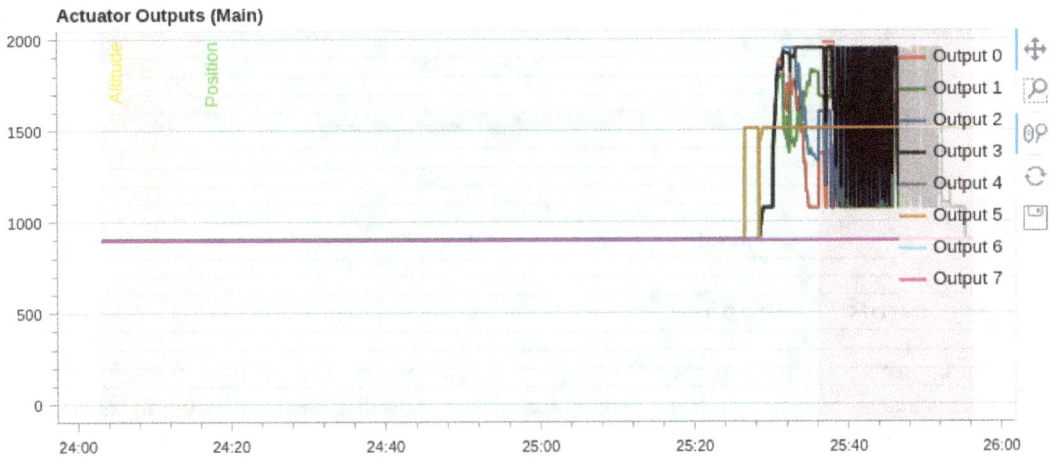

Figure 10.6 – Actuator output graph

Angular acceleration FFT graph

A **Fast Fourier Transform** (**FFT**) graph is used to visualize signals. In log analysis, this is used to convert vibration into amplitudes and display it as a frequency in the frequency domain. It is often used in time series graphs to determine which frequency components fall under which data time series. An angular acceleration FFT graph displays the frequency components of the drone's angular accelerations. It helps identify abnormal vibrations or oscillations in the drone's motion, which can be useful in detecting anomalies related to structural issues, sensor noise, or disturbances. Analyzing this graph can assist in finding anomalies and ensuring smoother and safer drone flights:

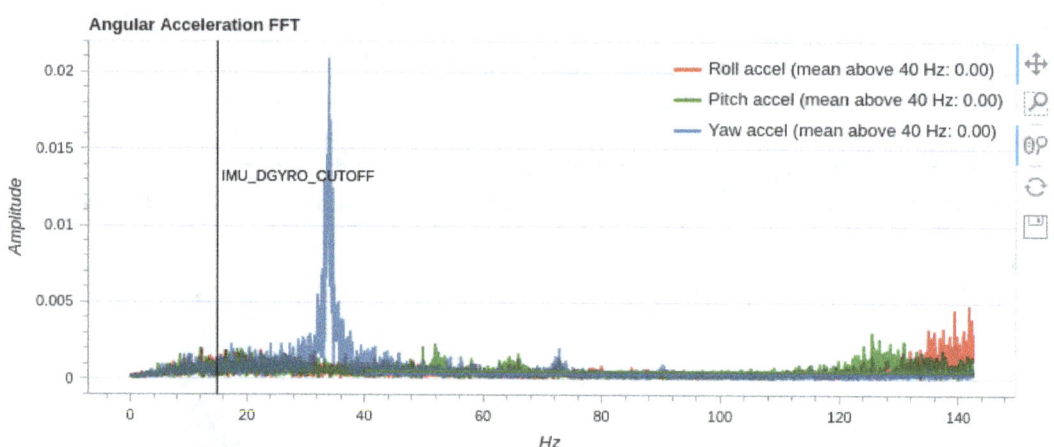

Figure 10.7 – Angular acceleration FFT graph

Actuator controls FFT graph

An actuator controls FFT graph displays the frequency components of control inputs to the drone's actuators. It's valuable for identifying irregular control inputs or oscillations, which can indicate issues with the FCS or control commands. Analyzing this graph helps spot anomalies and improves drone flight stability and performance:

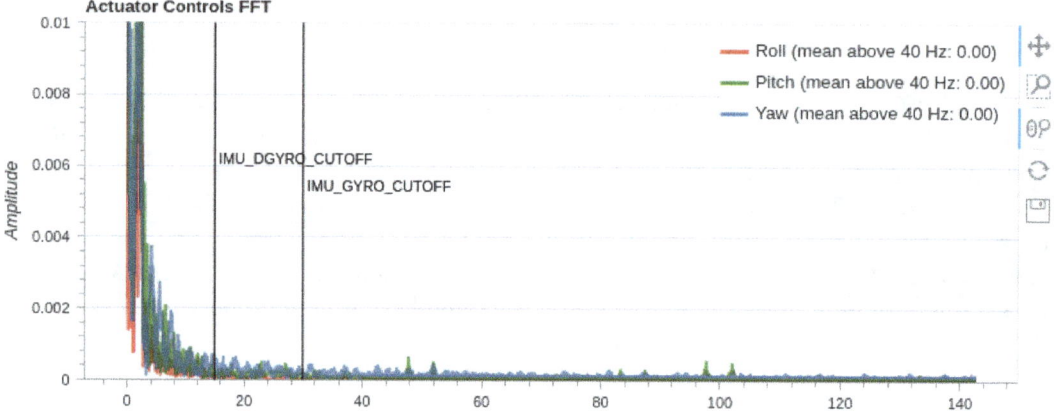

Figure 10.8 – Actuator controls FFT graph

Let's see some examples of good vibration graphs and bad vibration graphs.

The following graph is an excellent vibration graph where the amplitude for frequency of different Hz is minimal. It shows that vibrations are within an acceptable range, ensuring a stable and well-balanced flight. Deviations from this range suggest issues such as mechanical problems, unbalanced components, or flight instability, which can impact performance and safety:

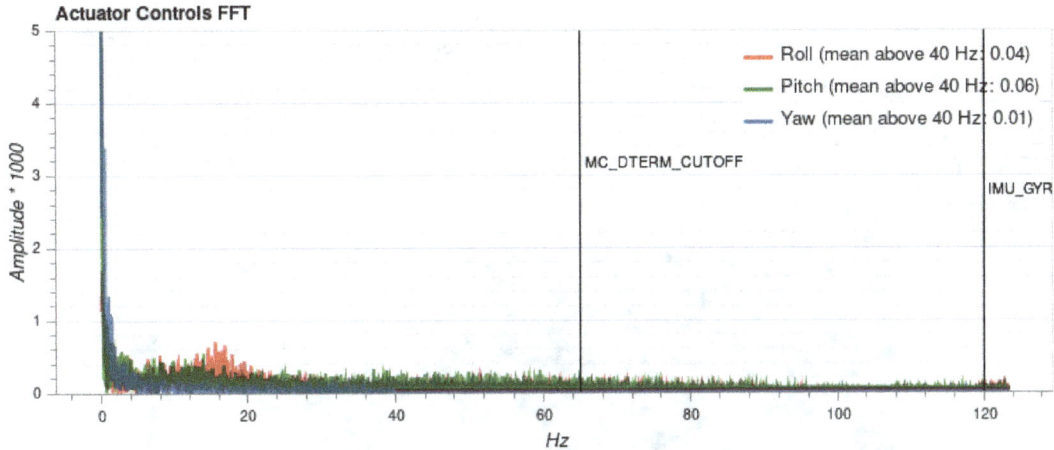

Figure 10.9 – Actuator controls FFT graph

The following graph exhibits frequent, high-amplitude spikes and consistently high values, signifying excessive and irregular vibrations during the flight. These irregular patterns and deviations from an acceptable range can impact the drone's stability and sensor data quality and pose a risk of damaging components. It is a clear indicator of issues affecting drone performance, hence the drone has a very poor vibration:

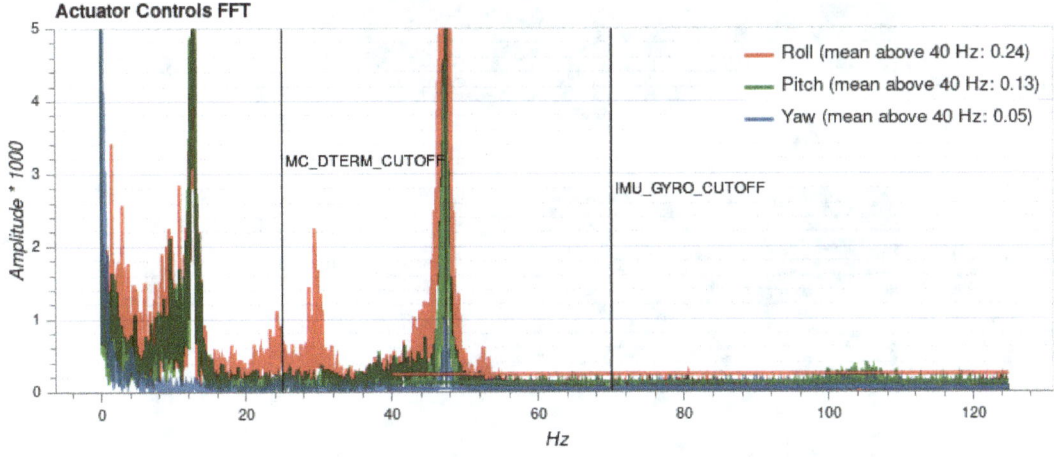

Figure 10.10 – Actuator controls FFT graph

Acceleration power spectral density graph

An acceleration **power spectral density** (**PSD**) graph is a 2D graphical representation that displays the frequency response of raw accelerometer data over time during a drone's flight. It provides insight into how the drone experiences acceleration at different frequencies and amplitudes. By analyzing this graph, we can gain a better understanding of the drone's motion patterns and vibrations, helping to diagnose issues related to mechanical problems, flight stability, and sensor anomalies. The more yellow the area is, the higher the frequency of responses.

The graph helps identify whether the drone's acceleration behavior is within acceptable limits, and any deviations or irregularities can signal problems that can be later rectified:

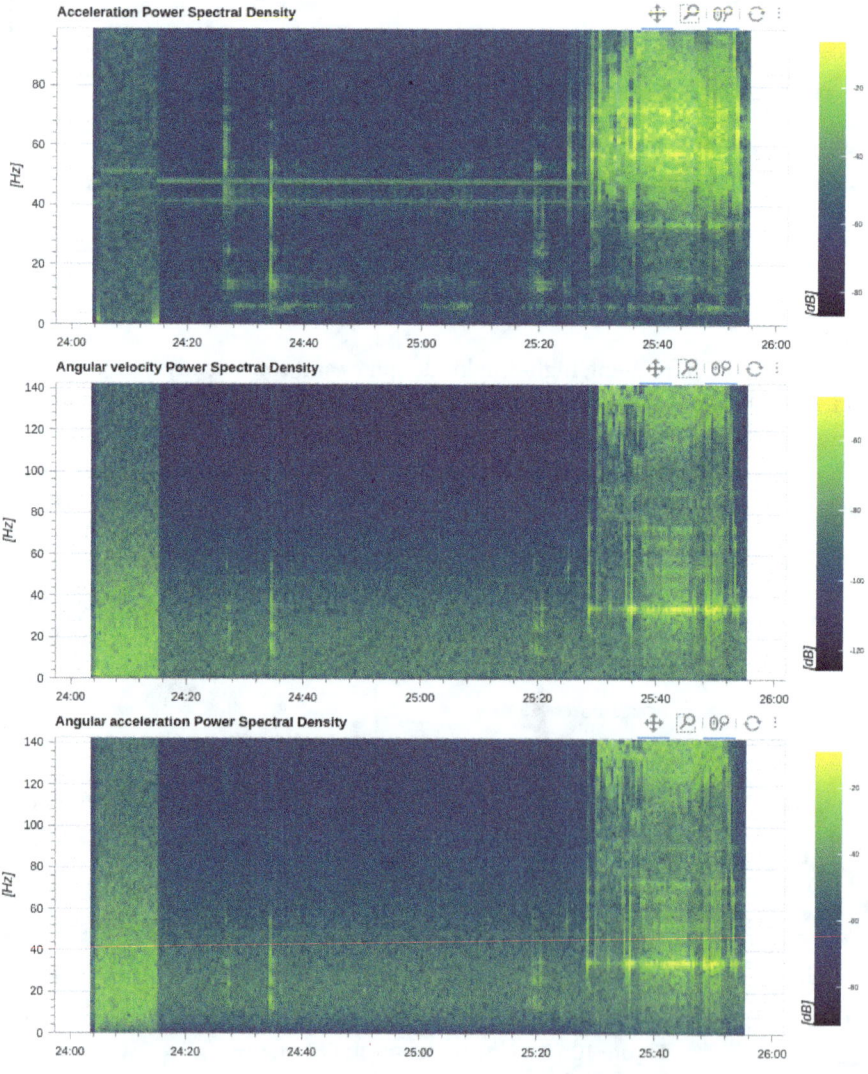

Figure 10.11 – Acceleration PSD graph

Power graph

A power graph displays the drone's power consumption trends during its flight. It helps to observe current and voltage values, which are critical for understanding battery performance and system efficiency. By analyzing this graph, we can monitor whether the drone remains within safe power limits, ensuring that it operates optimally and for the expected duration. Significant fluctuations or irregularities in current and voltage values can indicate potential issues that need attention to maintain safe and efficient drone operations:

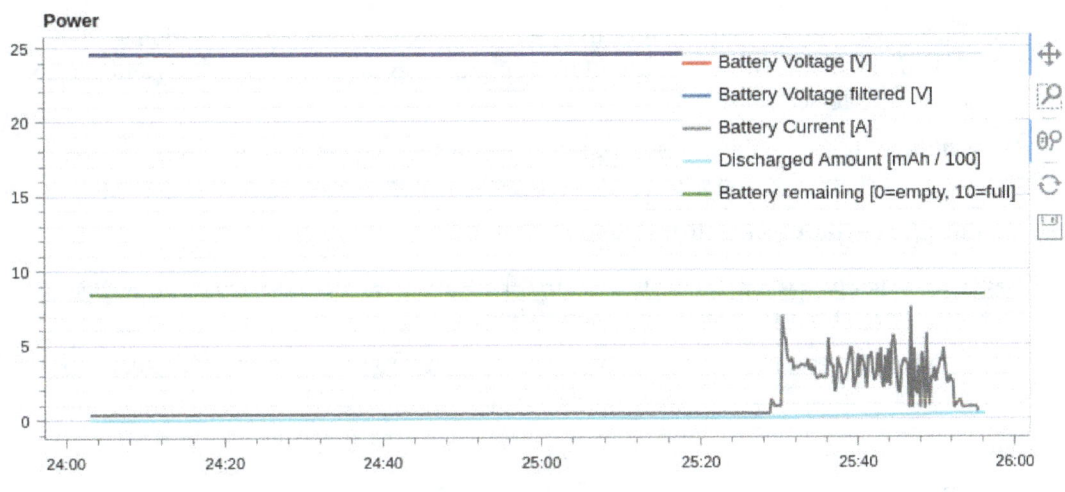

Figure 10.12 – Power graph

Understanding PID controllers and their uses

PID control is a fundamental and widely used method in the world of drones, serving as the backbone of flight stabilization and control systems. Drones rely on PID controllers to maintain stable flight, precisely follow flight paths, and respond to pilot commands. This control algorithm dynamically adjusts the drone's orientation and position by constantly measuring the error between the desired and actual states. In a nutshell, PIDs in drones play a pivotal role in ensuring smooth and controlled flight, making them an essential concept for both drone enthusiasts and professionals in the field.

PIDs are necessary for drones because they provide a mechanism for real-time adjustments to maintain stability, control orientation and position, respond to environmental factors, and ensure a smooth and safe flight. Here's a detailed explanation of why PIDs are necessary for drones and their primary uses:

- **Stabilization**:

    Use: PIDs are crucial for stabilizing the drone's flight. They continuously monitor the drone's orientation and make real-time adjustments to keep it level and maintain a steady position.

    Importance: Without effective stabilization, drones would be subject to uncontrollable movements, making them difficult to fly and posing safety risks.

- **Attitude control**:

    Use: PIDs help control the drone's attitude (orientation) by adjusting motor speeds and control surfaces. They ensure that the drone responds to pilot commands and maintains the desired roll, pitch, and yaw angles.

    Importance: Attitude control is essential for maneuvers, maintaining a straight flight path, and responding to pilot inputs accurately.

- **Position hold and waypoint navigation**:

    Use: PIDs enable drones to maintain a fixed position in GPS coordinates (position hold) and follow predefined flight paths (waypoint navigation). They adjust motor speeds and control surfaces to counteract external factors such as wind.

    Importance: Position hold and waypoint navigation are vital for applications such as aerial photography, surveying, and search and rescue, where precise positioning is necessary.

- **Response to environmental disturbances**:

    Use: PIDs help drones respond to environmental disturbances, such as wind or turbulence, by making immediate corrections to maintain stability.

    Importance: Environmental disturbances can impact a drone's flight. PIDs help counteract these effects to ensure smooth and controlled flight.

- **Smooth flight**:

    Use: PIDs contribute to smoother flight by reducing oscillations and overshooting. They provide gradual corrections to maintain a stable flight path.

    Importance: Smooth flight is essential for capturing high-quality images and videos, especially in aerial photography and videography.

- **Safety**:

    Use: PIDs play a critical role in ensuring safe drone operations. They prevent erratic movements and help avoid dangerous situations, reducing the risk of crashes.

    Importance: Safety is a paramount concern in drone operations, both for the drone and the people and property it interacts with.

- **Error correction**:

    Use: PIDs continuously monitor the error between the desired and actual drone state. They adjust control surfaces and motor speeds to minimize this error.

    Importance: Error correction is essential for maintaining precision in flight and meeting the operator's intentions.

Here's an explanation of what PIDs are and their role in drones:

- **Proportional** (*P*): The proportional component of the PID controller measures the current error between the desired and actual state of the drone. In the context of drones, this could involve the difference in orientation (roll, pitch, yaw) or position (latitude, longitude, altitude). The *P* term calculates how much and in which direction the drone should adjust to minimize the error. If the error is large, the correction applied by the proportional component is significant, and if it's small, the correction is minor.

    Use: The proportional term calculates a corrective action proportional to the error. If the drone deviates from the desired state, the *P* term applies a correction that is directly proportional to the magnitude of the error.

    Importance: The *P* term's primary role is to provide an immediate response to errors, helping the drone maintain the desired orientation, position, or flight path.

- **Integral** (*I*): The integral component considers the accumulation of past errors over time. It is responsible for addressing any long-term errors that the proportional component alone cannot fix. For drones, this can help correct slow, persistent disturbances, such as wind or minor imbalances in the drone's structure.

    Use: The *I* term addresses long-term or persistent errors. It helps correct slow, continuous disturbances or errors that the *P* term alone cannot resolve. For instance, it compensates for slight imbalances or biases in the drone's system.

    Importance: The *I* term is critical for fine-tuning and eliminating steady-state errors that may not be immediately addressed by the *P* term.

- **Derivative** (*D*): The derivative component accounts for the rate of change of error. It helps prevent overshooting or oscillations around the desired state. In drone control, the derivative component can reduce the impact of rapid changes in orientation or position, such as abrupt maneuvers.

    Use: The *D* term helps dampen the drone's response to rapid changes in error. It mitigates the potential for overshooting or oscillations that can occur during quick maneuvers or corrections.

    Importance: The *D* term is essential for stabilizing the drone and ensuring that it does not exhibit erratic behavior or excessive oscillations, especially during aggressive flight or when responding to sudden disturbances.

Methods to detect bad PIDs in drones and how to fix them

We need to know from continuous flights whether we need to tune our drone or not. Here is a list of observations that we can make from the flight behavior and what solutions we can utilize for them:

- **Excessive vibrations and oscillations**:

  Symptoms: If your quadcopter exhibits pronounced vibrations and oscillations, especially during hover or steady flight, it may indicate issues with the PID values. You might see rapid back-and-forth movements or jittery behavior.

  Solution: Check and adjust the P, I, and D values to reduce vibrations and achieve smoother flight.

- **Slow response to commands**:

  Symptoms: A quadcopter that responds slowly to pilot inputs, particularly during maneuvers or directional changes, may signal suboptimal PIDs.

  Solution: Increase the P values for more responsiveness. Adjusting the P term makes the quadcopter react more quickly to commands.

- **Overshooting and bouncing**:

  Symptoms: If the quadcopter overshoots its desired position and then bounces back, it could indicate that the P term is set too high.

  Solution: Decrease the P value until the overshooting stops while maintaining stability. Fine-tune it to achieve a balance.

- **Inaccurate altitude hold**:

  Symptoms: Altitude fluctuations during hovering or visible deviations in height may indicate the I term requires adjustment.

  Solution: Increase the I value to address altitude drift or decrease it if the drone tends to fluctuate. Find the point where the quadcopter maintains a consistent altitude.

- **Yaw drift**:

  Symptoms: Yaw drift occurs when the quadcopter's heading slowly changes without pilot input, often due to incorrect D-term tuning.

  Solution: Modify the D value to reduce yaw drift. Increasing D helps reduce oscillations and maintain heading stability.

Rate controllers in a drone and tuning PIDs

Drones use a sophisticated control system to maintain stable flight and respond to pilot commands or autonomous instructions. This control system includes various controllers, loops, and rate controllers. Tuning individual gains and parameters within these controllers, especially in PID controllers, is crucial for optimizing drone performance. Here's a detailed explanation of these components and the process for tuning gains and parameters:

- **Attitude controller**:

    Role: The attitude controller is responsible for maintaining the drone's orientation, which includes roll, pitch, and yaw angles. It ensures the drone remains level and responds to user inputs or autopilot commands.

    Architecture: The attitude controller is often implemented as a PID controller.

    The input to this controller consists of the desired attitude (from user or autonomous commands) and the current attitude measured by onboard sensors, such as accelerometers and gyroscopes.

    Gains: The primary gains for attitude control include P (Proportional), I (Integral), and D (Derivative) gains for roll, pitch, and yaw control. They are usually denoted as `Roll_P`, `Roll_I`, `Roll_D`, `Pitch_P`, `Pitch_I`, `Pitch_D`, `Yaw_P`, `Yaw_I`, and `Yaw_D`.

    Tuning gains: Adjust `Roll_P`, `Pitch_P`, and `Yaw_P` to control the drone's sensitivity to attitude changes. Higher values make the drone react more aggressively to level deviations.

    Tune `Roll_I`, `Pitch_I`, and `Yaw_I` to eliminate steady-state errors. Increasing these values helps maintain a precise attitude.

    Modify `Roll_D`, `Pitch_D`, and `Yaw_D` to reduce oscillations and overshooting during attitude changes. Increasing D dampens the response to rapid changes in attitude.

- **Rate controller**:

    Role: The rate controller focuses on stabilizing the drone by adjusting motor speeds to maintain specific angular rates (rotation speeds) around each axis (roll, pitch, and yaw).

    Architecture: As with the attitude controller, the rate controller often employs a PID structure.

    It receives inputs of the desired angular rates (from the attitude controller) and the actual angular rates measured by onboard gyroscopes.

    Gains: The rate controller includes P, I, and D gains for roll, pitch, and yaw rate control, which are typically denoted as `RollRate_P`, `RollRate_I`, `RollRate_D`, `PitchRate_P`, `PitchRate_I`, `PitchRate_D`, `YawRate_P`, `YawRate_I`, and `YawRate_D`.

    Tuning gains: Adjust `RollRate_P`, `PitchRate_P`, and `YawRate_P` to control the drone's sensitivity to angular rate changes. Higher values make the drone react more aggressively to changes in rotation.

    Tune `RollRate_I`, `PitchRate_I`, and `YawRate_I` to eliminate steady-state errors in angular rate control. Increasing these values helps maintain precise control.

    Modify `RollRate_D`, `PitchRate_D`, and `YawRate_D` to reduce oscillations and overshooting during rate changes. Increasing D helps dampen the response to rapid rate changes.

If gains are not tuned, they need to be tuned. The steps involved in the tuning process are as follows:

1. Start with the baseline values provided by your flight controller or manufacturer.
2. Make small, incremental adjustments to gains and parameters, changing only one value at a time.
3. Conduct test flights in a controlled environment to evaluate the effects of the adjustments.
4. Observe the drone's behavior in response to pilot inputs and environmental conditions.
5. Continuously review and fine-tune gains and parameters iteratively, considering the specific requirements of your drone and operating conditions.
6. Analyze flight logs and data to assess the drone's performance in more detail.
7. Engage with the drone community and forums for advice, share your PID settings, and learn from experienced pilots.
8. Prioritize safety during the tuning process and be prepared to take control or land the drone in case of unexpected behavior.

Tuning gains and parameters within attitude and rate controllers is a critical process for achieving stable and responsive flight. It allows you to optimize the drone's performance for your specific needs, whether it's for smooth aerial photography or precise acrobatic maneuvers.

Summary

So, in this chapter, we have learned a lot about the log analysis of drones, where they are stored, how they can be accessed, and how they can be interpreted. This is not just a one-day game; it requires an amount of experience and skills that will come over time to understand these log files' graphs and derive a conclusion from them. We have also studied PIDs and their importance and role in the smooth flying of quadcopters. We have also seen different rate controllers, how to tune individual PID gains, and how they impact the performance of the flight. We have also gone through different kinds of graphs and logs and learned how to analyze them. By studying all this, we got to know the basics and started tuning, which will be helpful for us in further tuning and analysis and achieving a smooth flight.

In the next chapter, we will study how we can implant different payloads onto a drone and how can we configure them.

11
Application-Based Drone Development

Earlier in this book, we learned about the different types of drones and the basic components that are required to build one. We also learned about the basic physics of a drone and how to achieve stable flight with the help of engineering and physics. Drone technology continues to evolve, creating boundless opportunities across various industries. Advancements in drones enable enhanced capabilities, from increased flight efficiency to better data collection and analysis. Upgrading and feature development drive innovation, fostering safer, more precise, and versatile drones. Learning about drone technology opens doors to numerous career paths, from aerial photography and cinematography to agriculture, infrastructure inspection, and environmental monitoring. Understanding drone advancements empowers individuals to harness these tools effectively, contributing to improved efficiency, data accuracy, and problem-solving in diverse fields, ultimately shaping the future of technology and industry. This will help you understand various aspects of how and what features can be implemented in the system.

In this chapter, we'll go through the following topics to understand the different applications for which a drone and its features are developed:

- Survey-based drones
- Agricultural spraying drones
- Arial delivery drones
- System redundancy

Technical requirements

The basic technical requirement for this chapter is a complete understanding of drone anatomy, as well as physics and flight time requirements. You should also have a clear understanding of the range, altitude, and flight operation environment.

Furthermore, you must have a solid understanding of the flight controller pins and outputs, alongside the usage of the various peripherals in synchronous. Finally, you must understand the ports and protocols that are required, as well as the configuration parameters, since these are equally important.

Survey-based drones

Survey-based drones are lightweight drones with high-resolution cameras as their basic payload. They are compatible with a variety of payload options. These drones are used to take images of the surface of the Earth and gather their geo-coordinates for proper mapping and processing. These images are then used to create a map, which is used by different stakeholders. The following figure provides an example of a survey drone:

Figure 11.1 – Example of a survey drone

Aerial surveying involves collecting geospatial data from above the Earth's surface using aerial platforms such as airplanes, helicopters, or, more commonly nowadays, drones. This method is utilized across various industries for mapping, surveillance, research, and analysis purposes.

Why aerial surveying is required

Aerial surveys have helped surveyors record accurate geospatial data efficiently. This saves them time and effort, which ultimately reduces the amount of expenses that are incurred. Here are some of the advantages of aerial surveys over traditional survey methods:

- **Accuracy and detail**: Aerial surveys provide highly accurate and detailed data, enabling precise mapping and analysis of large areas that might be challenging or time-consuming to cover from the ground.

- **Efficiency**: Compared to traditional ground surveys, aerial methods cover more ground in less time. This efficiency is crucial for infrastructure planning, environmental monitoring, agriculture, and disaster management.
- **Cost-effectiveness**: Aerial surveys can be more cost-effective for large-scale mapping projects as they reduce the need for extensive manpower and the resources required for ground surveys.
- **Accessibility**: Some areas, such as remote or hazardous terrains, might be inaccessible or risky for ground surveys. Aerial surveying provides a safe and efficient way to gather data from such locations.

The role of drones in aerial surveying

Drones have revolutionized aerial surveying due to their versatility, agility, and cost-effectiveness. Here's why they are useful:

- **Flexibility and maneuverability**: Drones can navigate diverse terrains and environments, flying at different altitudes and angles to capture detailed imagery and data.
- **High-resolution imaging**: Equipped with high-quality cameras and sensors, drones can capture high-resolution images and collect various data types, including photogrammetry, **Light Detection and Ranging** (**LiDAR**), and thermal imaging.
- **Cost-efficiency**: Drones are more affordable to operate than traditional aircraft, making aerial surveying accessible to smaller businesses and projects
- **Real-time data**: Drones can provide real-time data during flights, allowing for immediate analysis and adjustments to survey plans
- **Safety**: Using drones for aerial surveying minimizes the risks associated with human involvement in potentially hazardous or inaccessible areas

Overall, drones play a significant role in modern aerial surveying by offering a cost-effective, efficient, and accurate means of collecting geospatial data, making them invaluable tools across industries such as agriculture, construction, environmental conservation, and infrastructure development.

Payloads used in survey missions

In aerial survey missions, drones utilize various types of payloads tailored to specific data collection needs. These payloads encompass a range of sensors and equipment designed to capture diverse types of information. Here are some common types:

- **RGB cameras**: These cameras capture standard visible light and are fundamental for basic aerial imagery. They're used for photogrammetry, creating orthomosaic images, and generating high-resolution maps.

- **Multispectral and hyperspectral cameras**: These specialized sensors capture data across multiple bands of the electromagnetic spectrum, beyond human vision. They're essential for agricultural surveys, vegetation analysis, and environmental monitoring since they detect plant health, soil composition, and crop stress. They are mainly used in crop surveys and soil health analysis.
- **Thermal imaging cameras**: These sensors capture heat signatures to identify temperature variations. They're used in various applications, including search and rescue, building inspections, and environmental studies.
- **LiDAR**: LiDAR systems emit laser pulses and measure the time it takes for the light to return, creating highly accurate 3D maps. They're used for terrain modeling, infrastructure assessment, and urban planning due to their precision in capturing elevation and structure details.
- **Gas and chemical sensors**: These sensors detect and measure specific gases or chemicals in the atmosphere. They're utilized in environmental studies, as well as to monitor air quality or detect leaks in industrial settings.
- **Magnetometers and gravimeters**: These payloads measure variations in magnetic fields or gravitational forces. They're crucial for geophysical surveys as they can identify subsurface features such as mineral deposits or geological structures.
- **Aerial LiDAR and bathymetric sensors**: Specialized sensors for aerial or underwater surveys, these are often used in coastal mapping, shoreline assessments, or underwater topography mapping.
- **Photogrammetry add-ons**: Attachments or add-ons for cameras or sensors improve the capabilities of photogrammetric applications, enhancing accuracy and image quality.

Each type of payload serves specific purposes, enabling drones to gather diverse data types that are essential for different industries and applications. The choice of payload depends on the objectives of the survey mission and the type of information needed for analysis and decision-making.

Integrating the payload with the drone

Integrating a camera payload with a drone's flight controller involves ensuring seamless communication and synchronization between the camera and the drone's onboard systems. Here's an overview of the integration process and the relevant settings:

- **Physical integration**:
  - **Mounting**: Securely attach the camera payload to the drone using compatible mounting mechanisms, ensuring stability and proper alignment. Make sure the camera is directly pointing downward.
  - **Connections**: Establish physical connections between the camera and the drone's flight controller, power supply, and communication ports. Ensure proper cabling and compatibility between the devices. Make sure you study the payload manual and the steps to integrate the flight controller with the camera.

- **Flight controller configuration**:

    - **Camera triggering**: Set up the flight controller so that it triggers the camera at specific intervals or waypoints during the flight mission. This configuration ensures synchronized image capture according to the survey plan.

    - **Communication protocols**: Configure communication protocols between the flight controller and the camera payload. This includes settings for triggering signals, data transmission, and control commands.

    - **Power management**: Ensure the flight controller manages power distribution to the camera payload, providing consistent and adequate power during the mission.

    - **Gimbal control (if applicable)**: For gimballed cameras, configure the flight controller to control the gimbal's movement and stabilization, allowing the camera to maintain a steady position for image capture.

- **Camera settings**:

    - **Geotagging**: Enable geotagging on the camera to embed GPS coordinates into each captured image, allowing for precise georeferencing during post-processing.

    - **Triggering interface**: Ensure the camera is compatible with the flight controller's triggering interface. Set the camera so that it responds to trigger signals that are sent by the flight controller at specified waypoints or intervals.

    - **Exposure and capture settings**: Adjust camera settings such as exposure, shutter speed, ISO, and image format so that they match the survey requirements and environmental conditions.

- **Testing and calibration**:

    - **Pre-flight testing**: Conduct thorough pre-flight tests to ensure the integration and synchronization between the camera and flight controller are functioning correctly. Check triggering, image capture, and data transmission.

    - **Calibration**: Calibrate the camera payload and flight controller if required, ensuring accurate image capture, geotagging, and synchronization during the flight mission.

By effectively integrating the camera payload with the flight controller and configuring relevant settings, the drone can capture images precisely according to the survey plan, ensuring accurate and valuable data collection for aerial surveying missions.

The Real-Time Kinematics (RTK) and Post-Processed Kinematics (PPK) positioning systems

RTK and PPK are two advanced positioning technologies that are used in aerial survey drones to enhance the accuracy of geospatial data collection:

- **RTK**: RTK is a satellite navigation technique that provides real-time, centimeter-level accuracy in positioning data. It relies on a constant connection to a reference station or base station on the ground, which transmits correction signals to the drone in flight. The drone's onboard RTK receiver compares these correction signals with its GPS signals, enabling precise positioning in real time. RTK technology allows for highly accurate surveying and mapping without the need for post-processing, making it ideal for applications requiring immediate and precise results, such as precision agriculture or construction site monitoring.

- **PPK**: PPK technology also delivers high-precision positioning but operates slightly differently. With PPK, the drone records raw GPS data during its flight. After the mission, this data is post-processed using ground-based reference stations or precise satellite positioning data. By comparing the recorded data with highly accurate reference data, PPK calculates corrected positioning information retroactively. PPK offers similar centimeter-level accuracy as RTK but allows for more flexibility in data processing and eliminates the need for a continuous real-time connection to a ground station during the flight. This method is commonly used in surveying missions, where real-time data transmission might be challenging, such as in remote or inaccessible areas.

Mission planning for a survey mission

An effective flight plan for an aerial survey mission is crucial to ensure comprehensive coverage, accurate data collection, and safe operation of the drone. Here's an outline of the key components of an effective flight plan:

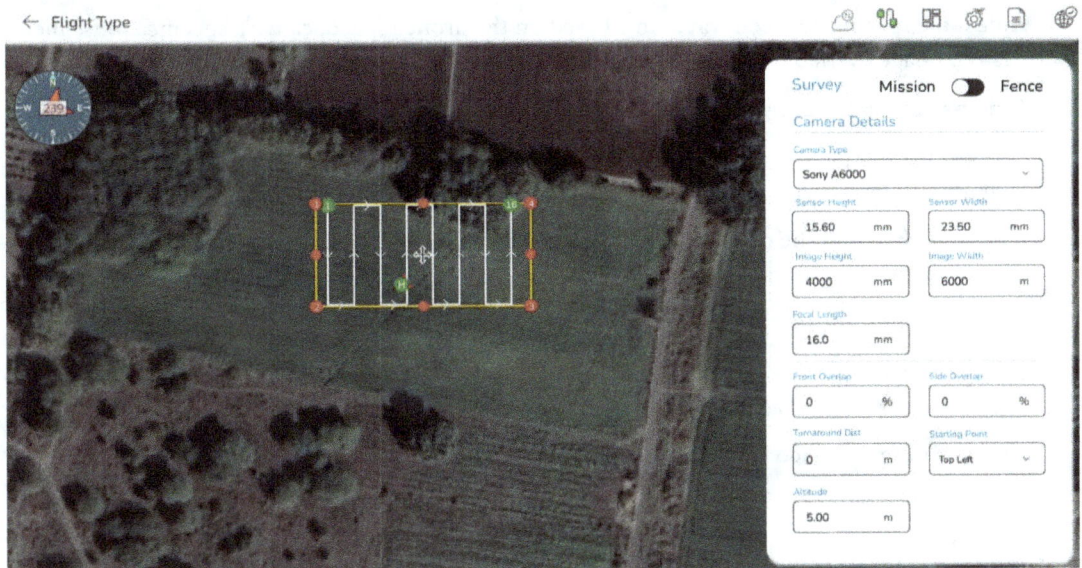

Figure 11.2 – Planning a survey mission in AeroGCS

The following are the key steps a pilot needs to follow before surveying an area with the survey drone:

1. **Define the survey area**:

    Clearly define the boundaries of the area to be surveyed using mapping software or a GCS.

    Take into account any obstacles, no-fly zones, or specific features within the survey area.

2. **Determine the altitude and resolution**:

    Calculate the optimal flight altitude based on the **Ground Sampling Distance (GSD)** required for the survey. The GSD determines the level of detail captured.

    In a survey, GSD refers to the physical size of one pixel on an image captured by a camera on the drone. It reflects the physical distance on the ground that each pixel corresponds to. A smaller GSD indicates higher spatial resolution, allowing for more detailed and accurate mapping and analysis in surveying applications.

    Consider the camera specifications, survey objectives, and terrain variations to set the appropriate altitude.

3. **Waypoint planning**:

    Use the GCS to plan the drone's flight path using waypoints. Ensure sufficient overlap between images (typically 60% to 80% in both directions) for accurate mapping.

    Plan the waypoints so that they cover the entire survey area while maintaining consistent altitude and speed.

Adjust the spacing between waypoints based on the drone's speed, camera specifications, and desired image overlap.

4. **Orientation and overlap**:

    Plan the drone's flight path so that it captures images with both forward and side overlaps to facilitate accurate image stitching and 3D reconstruction.

    Consider changing the orientation of the drone for cross-grid flights to maximize coverage and minimize distortion.

5. **Safety considerations**:

    Incorporate safety measures into the flight plan, including emergency landing spots, return-to-home points, and obstacle avoidance.

    Check for any airspace restrictions, environmental conditions, or potential hazards in the survey area.

6. **Test and validation**:

    Simulate the planned flight path in the GCS software to ensure complete coverage, adequate overlap, and adherence to safety protocols.

    Verify that the flight plan aligns with the survey objectives and desired image quality.

7. **Mission execution**:

    Execute the planned flight mission, ensuring the drone follows the predefined waypoints and captures images according to the set parameters.

    Monitor the flight to address any unforeseen issues and ensure data collection proceeds as planned.

    An effective flight plan is essential for achieving the survey's objectives, collecting high-quality data, and ensuring the success of the aerial survey mission. Adjustments and refinements to the plan based on real-time conditions and data evaluation contribute to the overall effectiveness of the survey.

Image stitching and post-processing

Image stitching is a process that involves combining multiple individual images into a single, seamless, larger image or panorama. In the context of aerial surveying, image stitching is crucial for creating orthomosaic images, which are high-resolution, geometrically corrected images of the Earth's surface. This process aligns and blends overlapping images to generate a comprehensive and accurate representation of the surveyed area.

We also need to understand that a trade-off exists between accuracy and making an image visually appealing since post-processing techniques can introduce distortions that will impact the accuracy of the stitched image. The important part is to produce visually good results without compromising the precision needed for reliable aerial survey data.

Several software options are commonly used for image stitching in aerial surveying:

- **Pix4D (www.pix4d.com):**

    Pix4D is a widely used photogrammetry software that offers advanced image processing capabilities for creating orthomosaics, 3D models, and point clouds. It uses **Structure-from-Motion (SfM)** algorithms to stitch images and produce accurate georeferenced maps.

- **Agisoft Metashape (formerly PhotoScan) (www.agisoft.com):**

    Agisoft Metashape is another popular photogrammetry software that's used for processing aerial imagery. It generates high-quality orthomosaic images and **Digital Surface Models (DSMs)** by aligning and blending images.

- **DroneDeploy (www.dronedeploy.com):**

    DroneDeploy is a cloud-based platform that offers image stitching and mapping functionalities for drone data. It automates the image processing workflow and creates orthomosaic images and 3D models for various industries.

These software applications utilize advanced algorithms to analyze and align overlapping images, correct distortions, adjust colors and exposure, and seamlessly blend them. They often allow for additional processing steps such as georeferencing, terrain modeling, and extracting various data layers.

Final deliverables

Aerial surveys produce various outputs that are valuable for different industries and applications. These outputs are derived from the data collected by drones equipped with specialized sensors during survey missions. Here are some key outputs:

- **Orthomosaic maps**: High-resolution, georeferenced maps that are created by stitching together multiple images captured during the survey. Orthomosaic maps provide a detailed and accurate representation of the surveyed area, which makes them useful for land planning, agriculture, infrastructure development, and environmental monitoring:

Figure 11.3 – Example of an orthomosaic image

- **Digital Elevation Models (DEMs) and DSMs**: DEMs represent the bare Earth's surface, while DSMs include surface features such as buildings and vegetation. These models offer elevation data, which is crucial for terrain analysis, flood modeling, urban planning, and infrastructure design:

Figure 11.4 – DSMs and DEMs

- **3D models**: Generated from aerial imagery, 3D models provide a detailed representation of the surveyed area in three dimensions. They're used in urban planning, construction, and archaeology, as well as for visualization purposes.
- **Vegetation indices**: Derived from multispectral or thermal imaging, vegetation indices indicate plant health, stress levels, or biomass. They're valuable in precision agriculture, forestry, and environmental studies.
- **Point clouds**: 3D representations of individual points collected by LiDAR or photogrammetry. Point clouds are used for detailed terrain modeling, volumetric analysis, and infrastructure inspections.
- **Change detection analysis**: Comparing images captured at different times allows changes in land use, infrastructure, or environmental conditions to be detected. This is beneficial for monitoring construction progress, environmental changes, and disaster assessment.
- **Environmental assessment data**: Data on soil composition, water quality, pollution levels, and habitat analysis obtained from aerial surveys aid in environmental impact assessments and conservation efforts.

With that, we've provided an in-depth analysis of survey-based drones. This will be helpful for new pilots to understand the basics of surveys. Next, we'll look at another application of drones – that is, in agriculture spraying.

Agricultural spraying drones

Agricultural spraying drones are specialized **Unmanned Aerial Vehicles** (**UAVs**) that are designed to apply fertilizers, pesticides, herbicides, and other agricultural inputs to crops in a precise and targeted manner. These drones have specific payloads and spraying systems that are optimized for efficient and controlled application:

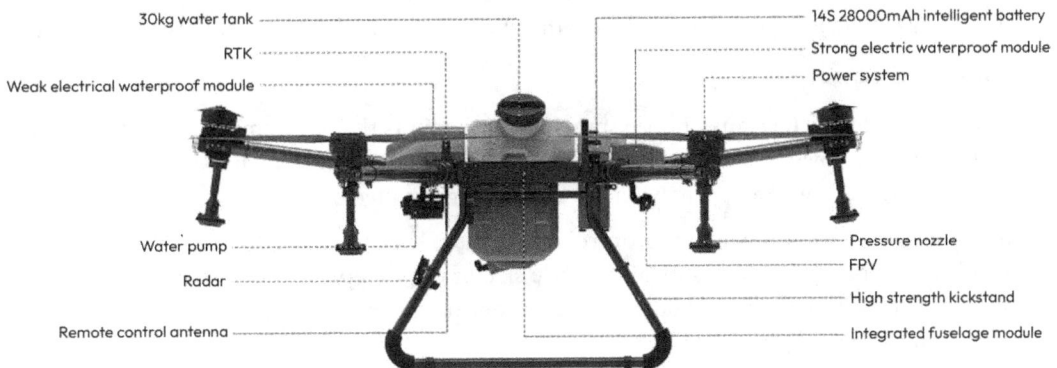

Figure 11.5 – Example of an agriculture drone

Let's take a brief look at payloads and spraying systems:

- **Spraying mechanisms**: Agricultural spraying drones are equipped with spraying systems that include pumps, nozzles, and tanks that are capable of carrying liquid payloads. These systems ensure controlled and even distribution of agricultural inputs over the crops.
- **Variable Rate Technology (VRT)**: Advanced spraying drones incorporate VRT, enabling spray rates to be adjusted based on crop conditions, allowing for targeted application, and minimizing input waste.
- **Payload capacity**: The payload capacity varies among different agricultural spraying drones, allowing for different volumes of agricultural inputs to be carried and applied during each flight. Some drones come with 5 L, 10 L, and 15 L capacity as the payload.
- **Navigation and precision**: These drones utilize GPS and precision flight control systems to navigate fields accurately, ensuring that the spraying is conducted in the right areas with minimal overlap or gaps. Effective GPS accuracy is required for this precision.

Spray tank integration with a drone

Nowadays, major spray gimbals are readily available and can be integrated with a drone. Integrating a 10 L spray tank with a drone and controlling the spray rate involves several steps to ensure the efficient and accurate application of agricultural inputs. Here's a guide:

Tank integration:

1. **Select the appropriate tank**: Choose a lightweight but durable tank that can securely hold 10 L of liquid. Consider materials that are compatible with agricultural chemicals and are not excessively heavy.
2. **Mounting system**: Design a mounting system on the drone's frame to secure the tank. Ensure the mounting system is balanced and stable so that it can maintain the drone's flight characteristics.

Spraying mechanism:

3. **Spraying system selection**: Choose spraying components that are compatible with the tank and the drone's power system. This includes pumps, nozzles, valves, and hoses that are suitable for handling the liquid and achieving the desired spray pattern.
4. **Spray rate control**: Implement a control mechanism that regulates the spray rate. This can involve adjusting pump pressure and nozzle size or even employing electronic flow control systems to modulate the flow of liquid. The rate control can be configured with the flight control PWM pins and configured by an RC.

Integration with the flight controller:

5. **Flight controller configuration**: Program the flight controller so that it can manage the spraying mechanism. Ensure that the flight controller can activate, deactivate, and regulate the spraying process during the flight.
6. **Interface setup**: The spray pump motor comes with an ESC, which controls the pump speed. Establish communication between the flight controller and the spraying system and control it

with any RC channel. This may involve configuring **Pulse Width Modulation** (**PWM**) or other control interfaces to relay commands from the flight controller to the spraying components.

Agriculture spraying flight planning

Flight planning for an agricultural spray drone is crucial for ensuring the precise and efficient application of agricultural inputs over fields. Here's an effective process for planning such missions:

Define mission parameters:

1. **Crop type and area**: Understand the type of crop and the size of the area to be sprayed
2. **Input requirements**: Determine the type and quantity of agricultural inputs (fertilizers, pesticides, and herbicides) needed for the specific crop and field conditions

    Assess environmental factors:

3. **Weather conditions**: Check weather forecasts for wind speed, direction, temperature, and humidity. Avoid flying in unfavorable conditions as this could affect spraying effectiveness or drone stability.
4. **Terrain and obstacles**: Analyze the terrain, elevation changes, and any obstacles (trees or power lines) that could impact the drone's flight path.

    Select flight path and parameters:

5. **Waypoint planning**: Use GCSs such as AeroGCS and Mission Planner to plan the drone's flight path with waypoints covering the entire field. Ensure adequate overlap between adjacent swaths for even spraying.
6. **Altitude and speed**: Determine the optimal flight altitude and speed while considering the spraying equipment and the desired spray coverage. Maintain a consistent altitude for uniform spraying.

    Spraying strategy:

7. **Spray rate and pattern**: Configure the spraying system so that it can achieve the desired spray rate and pattern for effective coverage. Adjust nozzle settings and flow rates based on crop type and input requirements.
8. **VRA**: Implement VRA if needed, adjusting the spray rate based on crop health or specific areas within the field.

Precautions and safety

Certain measures must be taken apart from normal UAVs since the chemicals that are used in the UAV can be dangerous to living beings and the environment. Also, the UAV will be large, so it must be handled carefully. Consider the following aspects:

- **Loss of control**: If control is lost, activate the *Return to Home* setting. Follow the drone's flight path and alert nearby personnel. In the case of a crash, use emergency procedures.

- **Use of safety equipment**: Wear appropriate **Personal Protective Equipment (PPE)** when handling drones with pesticides or after pesticides are used.
- **Dispose of containers**: Rinse the tank well after spraying. Puncture and safely dispose of the pesticide containers.
- **Mechanical inspections**: Inspect the UAV after drying for damage and record any issues.
- **Dispose of pesticide waste**: Triple-rinse the tank and properly dispose of waste to prevent people and animals from accessing the application area.

Aerial delivery drones

Drone deliveries represent an innovative approach to transportation that involves using UAVs or drones to deliver goods, packages, or items from one location to another. This method of delivery offers various advantages, including speed, accessibility, and the ability to reach remote or difficult-to-access areas. The basic features of a delivery drone are as follows:

- **Delivery-specific**: Aerial delivery drones are specialized for transporting goods, packages, or cargo from one point to another. Their design is more specific to the task of delivering items.
- **Enhanced payload capacity**: They are engineered to carry larger payloads, usually with a secure payload compartment or attachment mechanisms for transporting packages safely during flight.
- **Payload release mechanism**: These drones incorporate mechanisms to securely hold packages during flight and release them at specified locations. This involves payload release mechanisms or systems for controlled package drop-off.
- **Navigation and delivery precision**: There's an emphasis on precise navigation, accurate landing, and delivery capabilities at designated drop-off points.

Payload release mechanism

In a drone, a payload release mechanism is a device or system that's designed to safely deploy or release a payload, such as packages, equipment, or other items, during flight. It enables controlled and precise deployment of the payload at designated locations or times. The mechanism varies based on the drone's purpose and the specific requirements of the payload being released.

Let's look at the different types of payload release mechanisms:

- **Servo-based mechanism**: Utilizes servo motors to actuate and control the release mechanism. A servo is connected to a latch or mechanism that holds the payload. When activated, the servo moves, releasing the latch and allowing the payload to drop.

- **Electromagnetic release**: Uses an electromagnetic mechanism to hold the payload in place. When an electric current is applied or interrupted, it releases the magnetic hold, allowing the payload to detach.
- **Pneumatic/hydraulic release**: Operates using compressed air or hydraulic pressure to trigger the release. When pressurized or depressurized, the mechanism activates, enabling payload deployment.
- **Manual mechanism**: Involves a simpler manual release system where a latch or hook can be manually triggered by the operator. This is often used in basic DIY or custom-built drones.

How it works

The functionality of a payload release mechanism can vary based on its design and method of activation:

- **Activation mechanism**: The release can be triggered remotely by the drone operator through the flight controller using a designated switch or channel. This activation sends a signal to the mechanism to initiate the release process.
- **Secure holding**: The mechanism securely holds the payload in place during flight, ensuring it remains attached until the release command is given.
- **Controlled deployment**: Upon receiving the release signal, the mechanism activates, allowing the payload to be dropped or deployed from the drone. The release can be designed for precision, enabling accurate placement of the payload at specific locations.
- **Safety measures**: Reliable release mechanisms incorporate fail-safes to prevent accidental or premature deployments. These safety features ensure the payload is released only when intended, enhancing the overall safety and reliability of the operation.

Setting parameters for controlling a payload release mechanism via a flight controller involves configuring specific settings to manage the activation and behavior of the release mechanism. These parameters are customized within the flight controller's software or GCS to ensure precise and controlled operation. Here are some key parameters to consider:

- **Channel assignment**: Assign a dedicated channel to the transmitter (radio controller) and the flight controller for controlling the payload release mechanism. Ensure this channel is not used for other critical functions to avoid conflicts.
- **Output configuration**: Configure the designated channel as an output channel in the flight controller's software/GCS. Specify this output for controlling the payload release mechanism.
- **Channel mapping**: Map the transmitter's switch or control to the assigned output channel on the flight controller. This mapping allows the operator to trigger the release mechanism using the designated control on the transmitter.

- **Control activation setup**: Set the activation mode for the release mechanism. This might include choosing between a toggle switch, momentary switch, or proportional control based on the type of mechanism and desired control method.
- **Timing and delays**: Implement timing parameters to control the duration of the release signal. This could involve setting specific time intervals for activation, and ensuring the payload is released for the desired duration.
- **Control response**: Fine-tune the control response of the release mechanism. Adjust parameters related to servo travel limits, pulse width, or control signal range to ensure the mechanism responds accurately to the control inputs.

Hence, the drone will see some changes in its development and how it attaches the various release mechanisms. The current status of deliveries is in the pilot stage and the whole infrastructure has to be developed for this. But very soon, this would result in a full-fledged ecosystem that can deliver critical medical organs and medicines, along with groceries.

Safety regulations in the drone ecosystem

With different kinds of drones entering the market that are of different sizes and different levels of endurance to cater to the payload delivery demand in terms of weight and range, some regulations must be formed by governments when it comes to operating these drones. Most countries, such as the USA, India, and those in Europe, are coming up with drone policies and regulations that ensure the safety and smooth testing of these flights in LOS and BVLOS modes. The concepts of UTM and remote ID are also being introduced to manage drone traffic.

System redundancy

Redundancy in drones involves incorporating backup systems or duplicate components to ensure continued operation and mitigate the impact of failures in critical systems. This redundancy is crucial for enhancing safety, reliability, and fault tolerance, especially in scenarios where system failures can lead to catastrophic consequences. Here's an overview of various redundant components in a drone, particularly within the Pixhawk flight controller:

- **Redundant sensors**:
  - **GPS redundancy**: Pixhawk flight controllers often support redundant GPS modules. Utilizing two GPS modules from different manufacturers provides backup positioning information in case of GPS signal loss or failure in one module.
  - **IMU redundancy**: In Pixhawk, redundancy can be achieved with multiple IMUs. Dual or triple IMUs provide backup attitude and orientation data, ensuring stability and accurate flight control, even if one sensor fails.

- **Barometer redundancy**: Some Pixhawk-compatible drones incorporate redundant barometric pressure sensors to ensure accurate altitude measurements. Redundant barometers help maintain precise altitude data for safe operations.

- **Redundant power systems**:

  - **Dual battery setup**: Pixhawk-based drones can be configured with dual battery setups. This setup allows the drone to continue operating on one battery if the other fails or depletes unexpectedly, ensuring continuous power supply and extended flight time.

  - **Dual power inputs**: Some Pixhawk flight controllers support redundant power inputs. These systems enable the drone to switch seamlessly between power sources, such as different **Battery Eliminator Circuits** (**BECs**) or power modules, ensuring uninterrupted power supply to critical components.

- **Implementation within Pixhawk**:

  - **Configuration and setup**: You can utilize the Pixhawk flight controller's software, such as ArduPilot or PX4, to configure redundant sensors and power systems. You can also access the parameters related to redundancy setups within the flight controller's configuration.

  - **Hardware selection**: Choose Pixhawk-compatible hardware that supports redundancy setups. Select redundant sensors, GPS modules, power modules, and flight controllers that are designed to work in dual or triple configurations.

  - **Calibration and testing**: Calibrate the redundant sensors and power systems within the Pixhawk software. Ensure proper functionality and switch-over procedures. Conduct comprehensive testing to validate the redundancy setups in controlled environments.

  - **Fail-safe implementation**: Implement fail-safe protocols within the flight controller's settings to detect sensor or power failures. Configure the system to automatically switch to redundant components in case of primary system failure.

Redundancy in drones, particularly within the Pixhawk ecosystem, enhances system reliability, reduces the risk of catastrophic failures, and ensures continued operation, even in the event of component failures or signal losses, making it highly valuable in critical missions or scenarios.

Summary

In this chapter, we considered the development of drones for agriculture, aerial surveys, and aerial deliveries. This has showcased their immense potential to revolutionize various industries. Drones equipped with specialized payloads have enhanced crop monitoring, mapping, and data collection in agriculture, optimizing yields and resource management. In aerial surveys, these UAVs have enabled efficient and detailed mapping for urban planning, environmental monitoring, and infrastructure assessment. Moreover, aerial deliveries have introduced new possibilities for logistics and emergency response, offering swift and efficient transport solutions.

Moving forward, the next steps involve refining drone technologies and applications. This includes making further advancements in sensor technologies, payload capabilities, and autonomous navigation systems. Enhancing safety measures, such as anti-collision systems and secure communication, will be crucial for broader drone adoption. Moreover, integrating AI and machine learning for data analysis from drone-collected information will unlock deeper insights for decision-making in various sectors. Collaboration among regulators, developers, and industries is essential to establish standardized practices and regulations, fostering responsible and widespread drone use in diverse fields. Ultimately, continuous innovation and collaboration will propel the evolution of drones, unlocking their full potential for societal and industrial transformation.

12
Developing a Custom Survey Drone

We are coming to the end of this book. We have gone through various concepts relating to drones. In previous chapters, we studied different components and how they are assembled. In this chapter, we will learn how to develop a custom drone for aerial surveys. We will conceptualize the drone and configure it with the specific payload that would best fit for survey purposes.

Previously, we saw that a drone is a platform that carries a payload for a specific purpose. In this chapter, we'll analyze the requirements of drones for survey purposes, conceptualize a drone based on those requirements, and then build one. We will also go through the basic requirements of conducting a survey using a drone. Aerial surveys are a frequent application for many different kinds of drones as with drones, aerial surveys take less time and require minimum effort to produce accurate results that are helpful to stakeholders.

In this chapter, we're going to cover the following main topics:

- Geospatial surveys
- Setting up the requirements for a survey drone
- Selecting the drone's components
- Drone configurations

Technical requirements

There are no technical requirements as such for this chapter. The contents of the previous chapters will be enough to help you understand the concepts that will be covered here. However, prior experience in soldering and working with circuits is preferred. If you want to design your own airframe, then having design experience in SolidWorks or Catia is required.

Having a working knowledge of how to assemble and place components would be a bonus.

Geospatial surveys

A geospatial survey is a systematic and organized way of collecting, analyzing, and interpreting geospatial data relating to the surface of the Earth. This approach of surveying employs various technologies to capture, process, and present information about the physical, natural, and man-made features of the Earth's surface in the form of geographic coordinates. Geospatial surveys employ a wide range of techniques, which include satellite imagery, aerial photography, ground surveys with GPS and LiDAR, and advanced spatial data analysis using **Geographic Information Systems (GISs)**. The primary objective of a geospatial survey is to gather accurate and up-to-date information about the Earth's surface to make informed decisions.

The use of geospatial surveys in the industry

The use of geospatial surveys is multifaceted and spans various industries and applications. One significant application is in urban planning, where geospatial surveys aid in land-use analysis, infrastructure planning, and zoning assessments. Environmental monitoring is another critical domain as it utilizes geospatial data to observe changes in ecosystems, monitor deforestation, assess water quality, and respond to natural disasters effectively. In agriculture, geospatial surveys contribute to precision farming by providing insights into crop health, monitoring vegetation, and predicting yield. Additionally, geospatial surveys play a pivotal role in disaster management, transportation planning, natural resource management, and scientific research. By providing a detailed and accurate representation of the Earth's surface, geospatial surveys empower decision-makers with the spatial intelligence needed to address complex challenges and make well-informed choices across various sectors.

Setting up the requirements for a survey drone

We have already studied aerial surveys, their advantages, and the types of sensors that are used, so we won't dive into the details here. Instead, we will jump into the main features a drone should have for surveys.

High-quality imaging sensors

High-resolution cameras or specialized sensors, such as multispectral or LiDAR for capturing detailed and accurate survey data, are the first requirement. We should consider the types of surveys for which different sensors are required, such as RGB cameras and LiDAR, and also consider their weight as this will impact the total *all-up weight* and performance of the drone.

GPS

We will need precise GPS capabilities to georeference survey data and ensure accurate mapping is required since geo-tagged images are the prime requirement for surveys. Also, a precise GPS is required for autonomous drone flight. A GPS with RTK capabilities would be a better option in the long run.

Autonomous flight planning

We will need autonomous drone flight planning as this allows users to define survey areas, set waypoints, and execute predefined flight paths. This is a key requirement for capturing evenly distributed, high-quality images without any blur.

Long flight time

We will need an extended battery life to cover large survey areas in a single flight, reducing the need for frequent landings and battery changes. We will also need a good propulsion system to consume less battery and provide a longer flight time. For our survey drone, we will consider having a flight time of a minimum of 60 minutes so that we can cover large areas.

Real-time monitoring and telemetry

It's always great to have live video feeds or telemetry data for real-time monitoring of the ongoing survey missions, enabling adjustments to flight parameters if required. This will also help in monitoring the drone system's health on long flights.

Durability and weather resistance

We will need robust construction and weather-resistant design to withstand environmental conditions, including wind, rain, and temperature variations.

Payload capacity

As discussed previously, we need adequate payload capacity to carry the required sensors or cameras for the specific surveying task. This could be a plug-and-play system. Having a larger payload capacity can give us the privilege of being able to deploy multiple payloads with different weight capacities.

All-up weight or gross weight

It is better to have an all-up weight as low as possible that will consume less battery and provide us with longer flight time, while also having less aerodynamic drag for smoother operation.

Selecting the drone's components

Building a drone involves selecting and assembling various mechanical and electronic components to create a functional and reliable aerial platform. In this section, we'll look at the key components of a drone and the considerations for selecting them. Everything right from the airframe to the latest flight controller has to be selected to make a perfect drone that can fulfill our surveying needs. Make sure that the drone is durable and complies with the local regulations.

Airframe

As we already know, the airframe is the skeleton of a rigid material that holds the electronics together in their desired area to make them function together as a proper system.

Here are the selection considerations for the airframe:

- **Material (carbon fiber, aluminum, and so on)**: Since a carbon fiber body proves to be good in terms of weight versus strength, we'll use a carbon fiber frame.
- **Size and weight capacity**: The airframe's size depends on the propellers we choose so that no two propellers touch each other during the flight. The weight should be 400 to 500 grams (about 1.1 lb) for a 6 kg all-up weight. Additional weight can be added for camera mounts, landing gear, and any cases.
- **Frame type**: We always have the option to choose between a quadcopter, a hexacopter, or a fixed-wing drone. Since we are studying multicopters, we will use a quadcopter. Choosing a hexacopter would add more weight to this application and consume more power. We can go ahead with a hexacopter if the weight of the payload is higher, let's say a LiDAR, and we need more thrust and stability in the drone. To optimize the system with minimum weight and power consumption, we'll be using a quadcopter airframe.
- **Foldable or rigid design for portability**: A rigid design would be preferred but it depends on what the user wants to go ahead with. A foldable arms design would reduce the size and be easy to transport.

Keeping the aforementioned requirements in mind, we can design and manufacture the airframe ourselves; however, that is a separate process. As UAV engineers, our role is to develop a successful optimum drone from the components available for a given application and be able to fly and debug it if any issues occur.

Let's look at a few examples of commercially available airframes, such as the Tarot RC Peeper (*Figure 12.1*) and Tarot TL4X001 (*Figure 12.2*), which can be a good fit for survey-based applications. However, you can select your own frame if you wish:

Selecting the drone's components 239

Figure 12.1 – Tarot RC Peeper

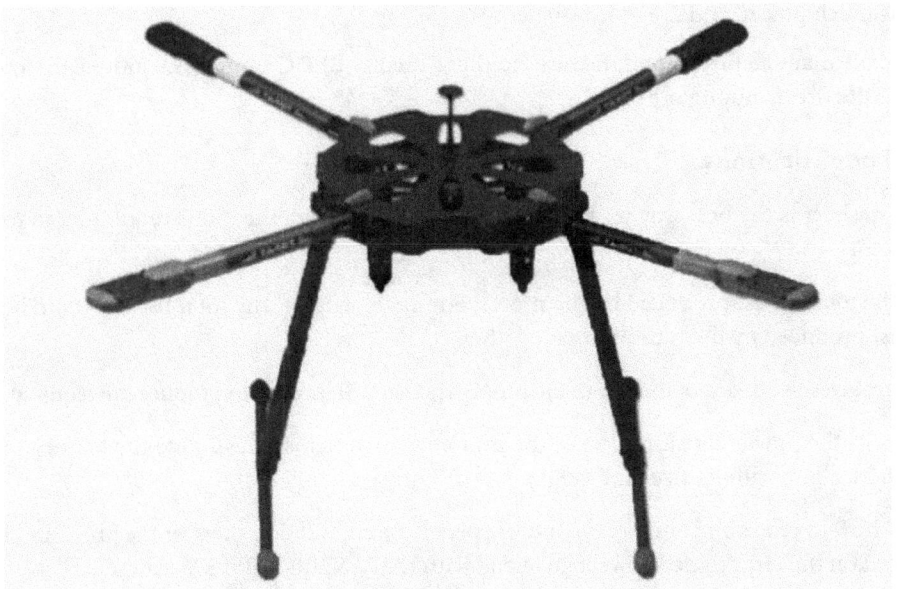

Figure 12.2 – Tarot TL4X001 specification

Propulsion system or power trail

The propulsion system is the most important part of the drone. It consists of the following parts:

- Motors
- Propellers
- Battery

- **Electronic Speed Controller (ESC)**
- **Power Distribution Board (PDB)**
- Power module

We must choose a propulsion system that meets the performance criteria of the drone.

Motor selection

The following are the selection considerations for motors:

- Type (brushed or brushless):
  - Drone technology uses BLDC motors to operate since they have a good life, don't consume as much power, and are more efficient
  - Many manufacturers manufacture excellent-quality BLDC motors, including T-Motor and KDE Direct, among others

Technical qualifications

A motor datasheet is the best way to understand the technical qualifications required and to select the best motor.

It reflects the total thrust produced by the motor. For a quadcopter, the total thrust would be equal to the thrust produced by the four motors together.

The datasheet gives us an idea of the size of the propellers that will be used to produce the required thrust.

It also gives us the power consumption of the motors, which helps us estimate the battery capacity that's required, along with its size and weight.

We learned how to read and compare the motor specifications and datasheet in the previous chapter. Hence, based on the requirements, we'll go ahead with the MN501-S 300 KV motor.

The following figure shows the datasheet of the MN501-S 300 KV motor. It contains the required parameters, such as weight, power, and resistance. We have chosen this motor due to these aspects:

KV	300	Rated Voltage(Lipo)	6-8S
Idle Current (18V)	1.1A	ESC Recommendation	AIR 40A
Peak Current (180s)	40A	Propeller Recommendation	20-22"
Max. Power (180s)	1000W	Motor Weight (Incl. Cable)	170g
Internal Resistance	63mΩ	Package Weight	350g

Figure 12.3 – Motor specification

Propeller selection

The manufacturer provides thrust bench data for the different propellers so that users can select the best propellers for their use cases – one that gives them the maximum thrust with minimum current consumption:

Type	Propeller	Throttle	Voltage (V)	Thrust (g)	Torque (N*m)	Current (A)	RPM	Power (W)	Efficiency (g/W)	Operating Temperature (°C)
MN5015 KV300	T-MOTOR P22*6.6	40%	24.34	1204	0.22	4.25	2333	103	11.64	HOT (Ambient Temperature :9.2°C)
		42%	24.34	1305	0.24	4.72	2422	115	11.36	
		44%	24.33	1421	0.27	5.28	2523	129	11.05	
		46%	24.33	1540	0.30	5.90	2627	144	10.73	
		48%	24.33	1666	0.33	6.62	2754	161	10.34	
		50%	24.33	1824	0.36	7.40	2859	180	10.14	
		52%	24.34	1980	0.40	8.23	2966	200	9.88	
		54%	24.32	2139	0.43	9.11	3276	222	9.65	
		56%	24.32	2266	0.46	9.94	3371	242	9.37	
		58%	24.32	2381	0.50	10.87	3463	264	9.00	
		60%	24.31	2530	0.54	11.83	3544	288	8.80	
		62%	24.31	2645	0.56	12.79	3639	311	8.51	
		64%	24.31	2803	0.60	13.98	3751	340	8.25	
		66%	24.31	2941	0.64	15.08	3835	367	8.02	
		68%	24.30	3008	0.67	16.11	3920	392	7.68	
		70%	24.30	3096	0.71	17.28	4090	420	7.38	
		75%	24.33	3507	0.80	20.38	4200	496	7.07	
		80%	24.32	4027	0.89	23.68	4478	576	6.99	
		90%	24.29	4655	1.06	31.11	4733	756	6.16	
		100%	24.27	5277	1.20	37.74	5017	916	5.76	

Figure 12.4 – Motor thrust data

The following are the selection considerations:

- Since the all-up weight of the drone is 6 kg, the thrust that's produced per propeller at 50% throttle should be at least 1,500 grams (about 3.31 lb).
- From this datasheet, we can see that thrust at 50% throttle is 1,824 grams, which is 324 grams (about 11.43 oz) above the requirement. This can compensate for the wind or any other requirements of high thrust.

The current thrust requirement is also fulfilled at 46%, which is producing 1,540 grams. This is equal to what we require, which is good. Hence, we can say that 22-inch propellers are an efficient combination for this motor.

Propeller material

The material of the propeller is important for its performance, efficiency, and durability. Some generally used materials for propellers are plastic, carbon fiber, and wood.

Plastic propellers are lightweight and affordable, making them suitable for beginners and casual drone enthusiasts, but they break often and may not provide the best performance.

Carbon fiber propellers are stronger and more rigid than plastic, which offers better performance and efficiency. They are also durable and resistant to damage, making them ideal for professional and high-performance drones.

Wooden propellers are less common but are flexible and durable. They are quieter than plastic and carbon fiber propellers, making them suitable for applications where noise is a concern.

The best material for a drone propeller depends on the specific requirements of the drone and its intended use. For high-performance drones, carbon fiber propellers are often preferred, while for casual use, plastic propellers may be more suitable.

Let's compare plastic, carbon fiber, and wooden drone propellers based on various factors:

Characteristic	Plastic Propeller	Carbon Fiber Propeller	Wooden Propeller
Weight	Light	Light	Light to moderate
Strength	Low	High	Moderate
Rigidity	Low	High	Low to moderate
Durability	Low	High	Moderate
Cost	Low	High	Moderate
Noise Level	Moderate	Low	Low
Performance	Basic	High	Moderate
Compatibility	Standard	High-end	Limited (mostly for small drones)
Impact Resistance	Poor	High	Moderate

Table 12.1 – Comparison of propeller materials

Hence, we can see that carbon fiber propellers offer better performance and durability and are a good choice for selection.

Airframe dimensions

To calculate the frame dimensions of a quadcopter when the length of the propellers is known, you can use the following formula:

$$Diagonal\ Length\ of\ Frame = Length\ of\ Propeller \times \sqrt{2}$$

This formula assumes that the propeller length is the distance from the center of the quadcopter to the tip of one of the propellers. The diagonal length of the frame is the distance between opposite motors (diagonally across the quadcopter).

Power and efficiency

We can see that the current consumption at 46% with the required thrust is 5.90A. Let's consider it as 6A for calculation purposes. The consumption by four motors is 24A. We'll keep the consumption by other peripherals set to 2A. Hence, the total current consumption is 26A.

Battery selection

The following are the selection considerations:

- The battery for this application should be one that have higher energy stored in less weight.
- **Lithium Polymer** (**LiPo**) batteries are used in drone applications as they can deliver higher current when required.
- Solid-state LiPo batteries are newer and are less weighty, with higher energy density.
- For 60 minutes of endurance, since the current consumption is 26A, we can go ahead with a battery with 6s 26 AH capacity solid state Li-Ion. However, we are bound by the battery capacity manufactured by the manufacturer. We must choose a battery capacity that is closest to our requirements.

Let's compare the use of solid-state Li-ion batteries with LiPo batteries in drones in terms of their power, C rating, and other electrical parameters, along with suitable applications for solid-state Li-ion batteries:

Characteristic	Solid-State Li-ion Battery	LiPo Battery
Energy Density	Moderate to high	High
Power Output	Moderate to high	High
C Rating	Moderate to high	High
Cycle Life	High	Moderate
Operating Temperature	Wider range	Limited
Charging Time	Moderate	Short
Cost	Higher	Lower
Performance	Good	Very good
Suitability	High-end drones, industrial applications	Consumer drones, hobbyist applications

Table 12.2 – Comparison of solid-state Li-ion and LiPo batteries

Solid-state Li-ion batteries are suitable for applications where safety, cycle life, and moderate to high power output are critical. They are ideal for high-end drones that are used in industrial applications, where performance and reliability are paramount.

There are different manufacturers across the globe that manufacture solid-state Li-ion batteries:

- Foxtech Diamond (www.foxtechfpv.com)
- Genx (www.foxtechfpv.com)

Based on the Foxtech battery available, the closest battery capacity that coincides with our requirement is 6s 27,000 mAh, which is 2,349 grams and 5c ratings. This is lighter compared to a standard LiPo battery. We can go ahead with any vendor with the same specifications but always choose a reputed vendor for the battery.

ESC selection

The following are the selection considerations:

- The criteria for ESC selection are quite simple. We know that each ESC is connected to a single motor and that there are four ESCs for four motors.
- From the motor datasheet, we calculated that the current consumption per motor at 46% is 6A. However, at 100%, it is 37.74A.
- To be safer and not let the ESCs burn, we will select the ESC that's rated at 40A and working on a 6s battery. Sometimes, the manufacturer recommends the ESC based on the motor selected and gives you the right propulsion system.

From *Figure 12.1*, we can see that for this motor and propeller combination, the recommended ESC by the manufacturer is 40A, hence we can go ahead with this ESC.

There are multiple ESC manufacturers, but the user can select whichever they want. Here are the names of some ESC manufacturers:

- T-Motor (www.store.tmotor.com)
- **Advanced Power Drives (APD)** (www.store.tmotor.com)
- Hobbywing (www.store.tmotor.com)

It is always advised to go through the complete documentation and datasheet for the ESC to know how to operate it.

Based on the selection consideration points mentioned here, we can choose the T-Motor Air 40 or APD 40F3 ESC:

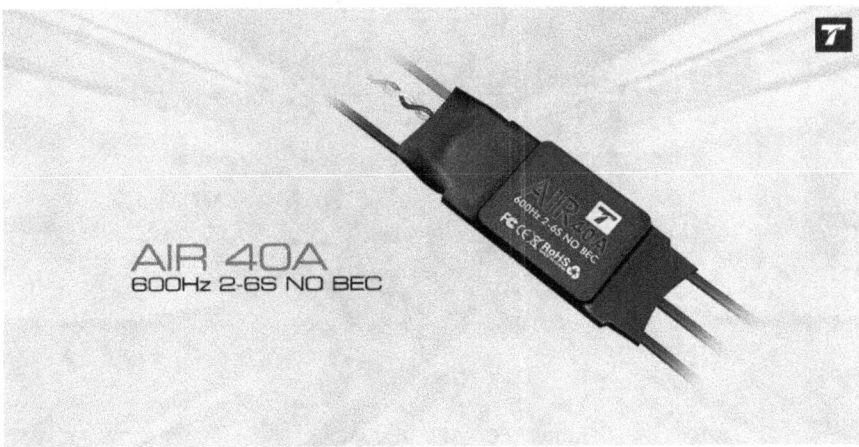

Figure 12.5 – T-Motor Air 40A ESC

PDB selection

The following are the selection considerations:

- We have seen that the maximum current consumed by the motor at 100% throttle is 37.5A. Hence, for all four motors, the current that will be consumed will be 149.6A.
- The peripheral devices, such as cameras and telemetry, would consume current based on their power requirements. Add an extra 5A for them.

  By considering these aspects and taking the maximum current as a buffer, we must use a PDB that can sustain a maximum current of 200A passing through it.

- Some additional ports at 5V and 12V would be an added advantage for powering the peripherals.

Here are some PDB manufacturers you can look into:

- Holybro (www.holybro.com)
- APD (https://powerdrives.net/)
- CUAV (www.cuav.net)

Please go through the datasheet for the PDB and check its polarity before connecting. It is always advised to check the voltage/continuity using a multimeter before powering the circuit.

Based on the selection criteria, we can go ahead with the APD PDB360(X), which has 360A of current-handling capacity and has stepped-down 5V and 12V supplies for the peripherals:

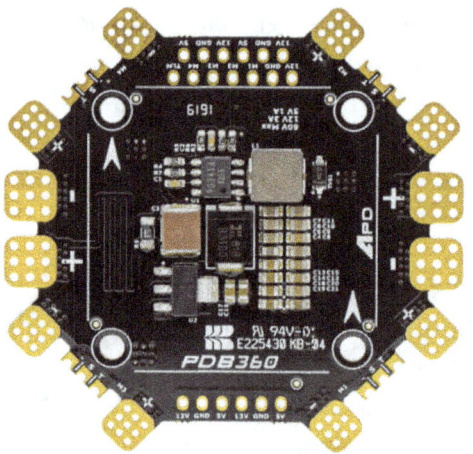

Figure 12.6 – APD PDB360 C

Autopilot selection

The following are the selection considerations:

- Compatibility with the frame and ESCs
- Sensor types (gyroscope, accelerometer, and magnetometer)
- Sensor redundancy and flight stack support
- Integration with GPS and other navigation systems
- Programming capabilities and community support

We have studied various flight controllers in the previous chapters. Here are some other examples of flight controllers that can be used:

- Cube Orange
- Pixhawk 6x
- CUAV X7/X7+

Let's compare the Pixhawk Cube Orange, Holybro Pixhawk 6X, and CUAV X7 drone flight controllers:

Feature	Pixhawk Cube Orange	Holybro Pixhawk 6X	CUAV X7
Processor	STM32H7	STM32H7	STM32H7
Sensors	IMU, barometer, magnetometer	IMU, barometer, magnetometer	IMU, barometer, magnetometer
GPS	Built-in	Built-in	Built-in
PWM Outputs	12	8	8
Serial Ports	8	5	6
CAN Bus	Yes	Yes	Yes
USB Ports	1 micro USB	1 USB-C	1 USB-C
Power Module	External	External	External
Dimensions	54 x 54 x 19 mm	44 x 84 x 12 mm	44 x 84 x 12 mm
Weight	45 g	31 g	33 g
Price	High	Moderate	Moderate
Supported Software	ArduPilot, PX4	ArduPilot, PX4	ArduPilot, PX4
Additional Features	ADS-B receiver	RGB LED, buzzer	Dual IMU, dual barometer

Table 12.3 – Comparison of different flight controllers

Based on its community support and functionality, we are going to use the Pixhawk Cube Orange flight controller. You can choose a different flight controller if you want.

Flight stack selection

The Pixhawk Cube Orange flight controller supports ArduPilot and the PX4 flight stack. For this build, and keeping the other components in mind, we'll go ahead with the PX4 flight stack. Complete details of the PX4 flight stack can be found at www.px4.io.

GPS module selection

The following are the selection considerations:

- GPS accuracy
- Compatibility with the flight controller
- Connection protocols supported
- Various GNSS constellations supported

The following are some other manufacturers and models of GPS modules for drones:

- Here 3
- CUAV NEO V2/V2 Pro
- Holybro Pixhawk 4
- Holybro H-RTK

Let's compare these:

Feature	Here 3	CUAV NEO V2	Holybro Pixhawk 4	Holybro HRTK
GPS Chipset	u-blox M8P	u-blox M8N	u-blox M8N	u-blox M8P
GPS Channels	72	72	72	72
GNSS Support	GPS, GLONASS, Galileo, BeiDou	GPS, GLONASS, Galileo, BeiDou	GPS, GLONASS, Galileo, BeiDou	GPS, GLONASS, Galileo, BeiDou
Position Accuracy	2.5 cm (RTK), 30 cm (GPS)	2.5 m (GPS), 0.8 m (RTK)	2.5 m (GPS)	2.5 cm (RTK), 30 cm (GPS)
Update Rate	5 Hz	5 Hz	5 Hz	5 Hz
Compatibility	ArduPilot, PX4	ArduPilot, PX4	ArduPilot, PX4	ArduPilot, PX4
Weight	99 g	53 g	39 g	34 g
Dimensions	80 x 80 x 25 mm	50 x 50 x 15 mm	40 x 40 x 10 mm	45 x 45 x 10 mm
Price	High	Moderate	Moderate	High

Table 12.4 – Comparison of various GPS modules

Based on these details, we'll choose the Here 3 GPS module.

Camera or payload

The following are the selection considerations:

- Camera resolution and quality
- Gimbal compatibility for stabilized footage
- Payload capacity and integration with the flight controller
- Purpose (for example, aerial photography, mapping, surveillance, and so on)
- The weight of the module
- Triggering mechanism from the flight controller

The following table compares the Sony Alpha 6000, Sony RX100, Sony RX0, and GoPro drone survey cameras in terms of an aerial survey:

Feature	Sony Alpha 6000	Sony RX100	Sony RX0	GoPro
Sensor Size	APS-C	1-inch	1-inch	1/2.3-inch
Megapixels	24.3	20.1	15.3	12
Lens	Interchangeable	Fixed	Fixed	Fixed
ISO Range	100-25,600	125-12,800	125-12,800	100-6,400
Shutter Speed Range	1/4,000 to 30 sec.	1/2,000 to 30 sec.	1/32,000 to 1/4 sec.	1/8,000 to 1 sec.
Video Resolution	1080p, 4K	1080p, 4K	1080p, 4K	1080p, 4K
Image Stabilization	Yes (in-camera)	Yes (in-camera)	Yes (in-camera)	Yes (in-camera)
Weight	344 g (without lens)	299 g	110 g	116 g
Price	Moderate	High	High	Moderate
Application	Photography, videography	Photography, videography	Photography, videography	Photography, videography
Additional Features	Fast autofocus, Wi-Fi	Zeiss lens, Wi-Fi	Shockproof, waterproof	Waterproof, voice control

Table 12.5 – Comparison of survey cameras

Based on these details, we have selected the Sony Alpha 6000 for better results and photographs. At the moment, because of this choice, our drone is 468 grams. There are other options available, such as the Sony RX100.

Telemetry system and selection considerations

The following are the selection considerations:

- Bidirectional communication with the GCS
- Range and reliability
- Data transmission rate
- Integration with the flight controller
- Lightweight, easy to handle, and mobility

Based on the selection criteria, we would like to go ahead with the Herelink communication system, which offers live video transmission via a micro HDMI, offers a range of more than 10 km, and has integrated RC, telemetry, and pre-installed QGroundControl station software. Please go through the datasheet for the modules to understand their integration and usage:

Figure 12.7 – The Herelink transmitter and receiver

You can use the following table to estimate the total weight of the drone based on the components you've selected. The weights that are specified here are not exact and will change based on the components you've selected:

Sl.No	Component	Weight	Quantity	Total Weight
1.	Airframe (including camera mounts)	1,100 g	1	1,100 g
2.	Motors	170 g	4	680 g
3.	Propellers	56 g	4	224 g
4.	ESCs	26 g	4	104 g
5.	PDB	100 g	1	100 g
6.	Battery	2,350 g	1	2,350 g
7.	Power module	25 g	1	25 g
8.	Flight controller	75 g	1	75 g
9.	Camera	468 g	1	468 g
10.	Herelink receiver	100 g	1	100 g
11.	Miscellaneous (wires, soldering, and so on)	100 g	1	100 g
	Total weight			5,326 g

Table 12.6 – Table showing the weight distribution of the drone

Hence, we can calculate that the total weight of the drone will be 5.3 kg. These calculations have been done while keeping the 6 kg all-up weight in mind. We still have an additional 674 grams if we wish to add any additional payload or sensor.

Wiring and assembly

A quick wiring diagram can be found on the official ArduPilot website at `https://ardupilot.org/copter/_images/Cubepilot_ecosystem.jpg`.

We won't focus too much on assembly and configuration here since we covered this in previous chapters:

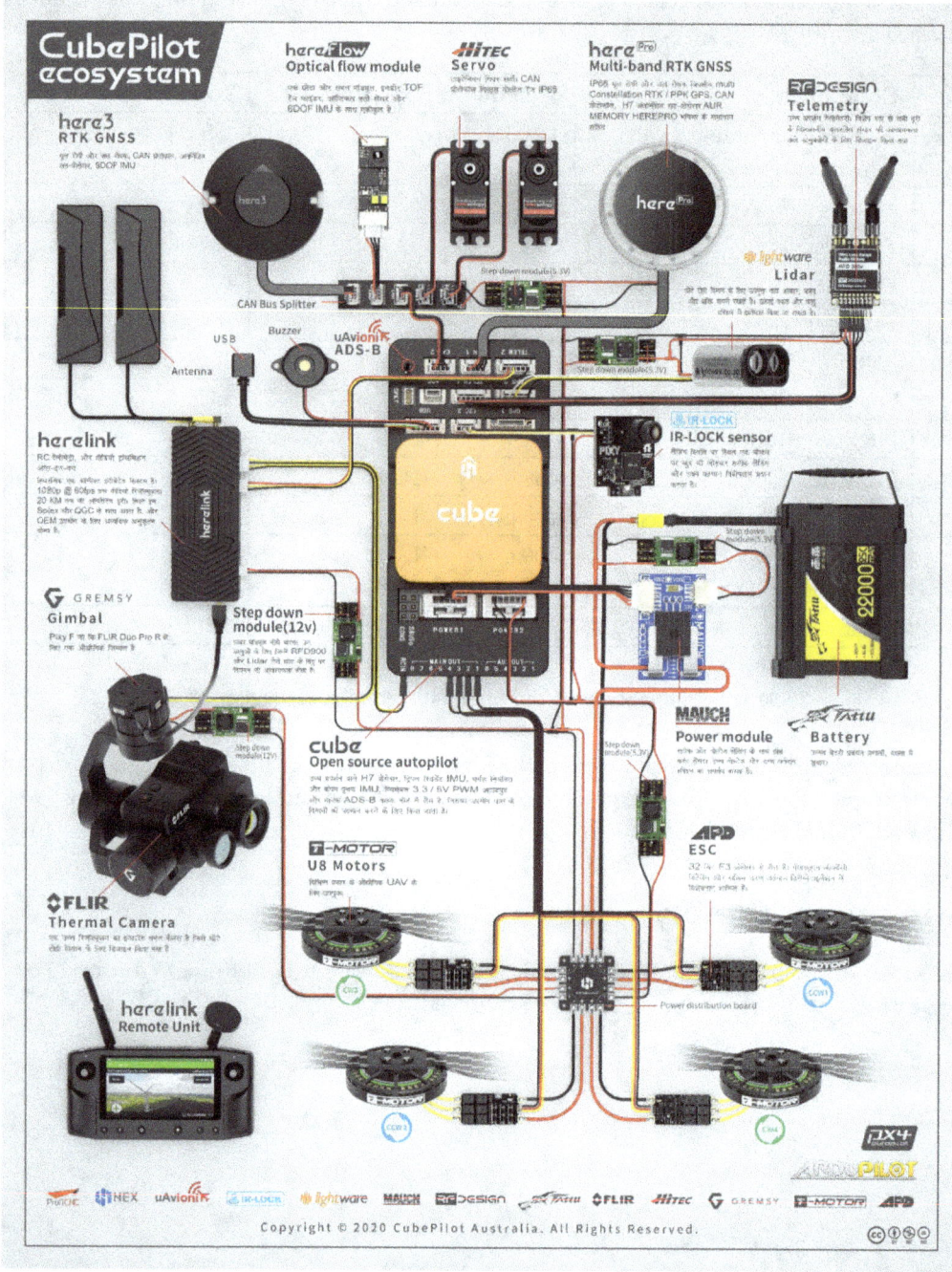

Figure 12.8 – Assembly diagram by ArduPilot

Drone configurations

After assembling the drone, we need to configure it so that it can fly. We'll need to conduct a series of steps to configure the drone based on AeroGCS. We studied the various options that are available in AeroGCS previously, so here, we'll quickly go through the process and select the right options for our use case (however, the same process can be configured with Qground Control as well using the Herelink GCS we have selected above):

1. Connect the drone via a USB cable.
2. Select an airframe:

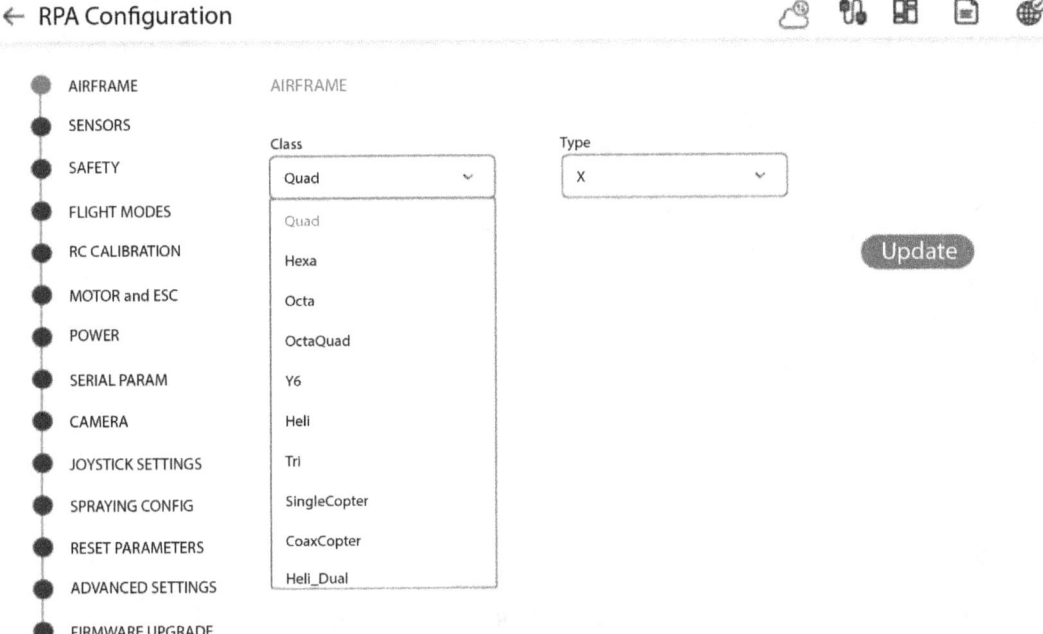

Figure 12.9 – Selecting an airframe in AeroGCS

Choose the following options:

- **Class: Quad**
- **Type: X** (considering the airframe is X-shaped)

Once you click **Update**, you'll see the ESCs getting PWM signals.

Next, we will calibrate the sensors.

3. First, we must calibrate the accelerometer. We learned about this previously. The **SENSORS** tab will help us with this:

    - Click **Calibrate**
    - Once you click **Calibrate**, the GCS will ask you to hold the drone in certain orientations for a few seconds
    - The autopilot will note the accelerometer readings of the different orientations and save them
    - Follow the instructions on the screen to calibrate the accelerometer:

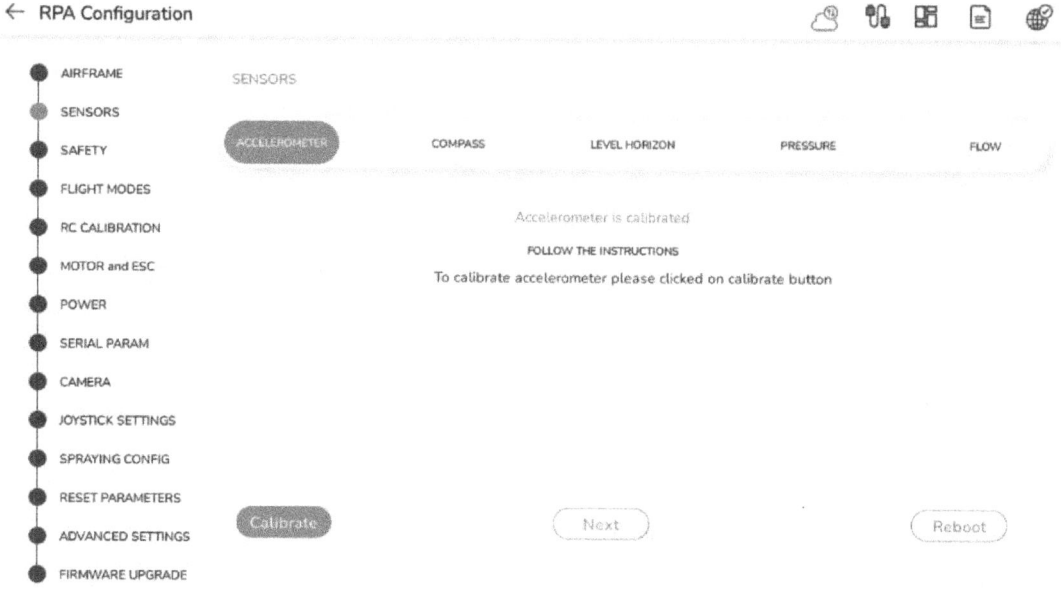

Figure 12.10 – The ACCELEROMETER window

4. Then, we have compass calibration. This is required for the autopilot to read the magnetic declination of the area. Click **Calibrate**, then rotate the drone on multiple axes. The progress bar will increase as the drone is rotated.

Once you've done this, the drone must be rebooted:

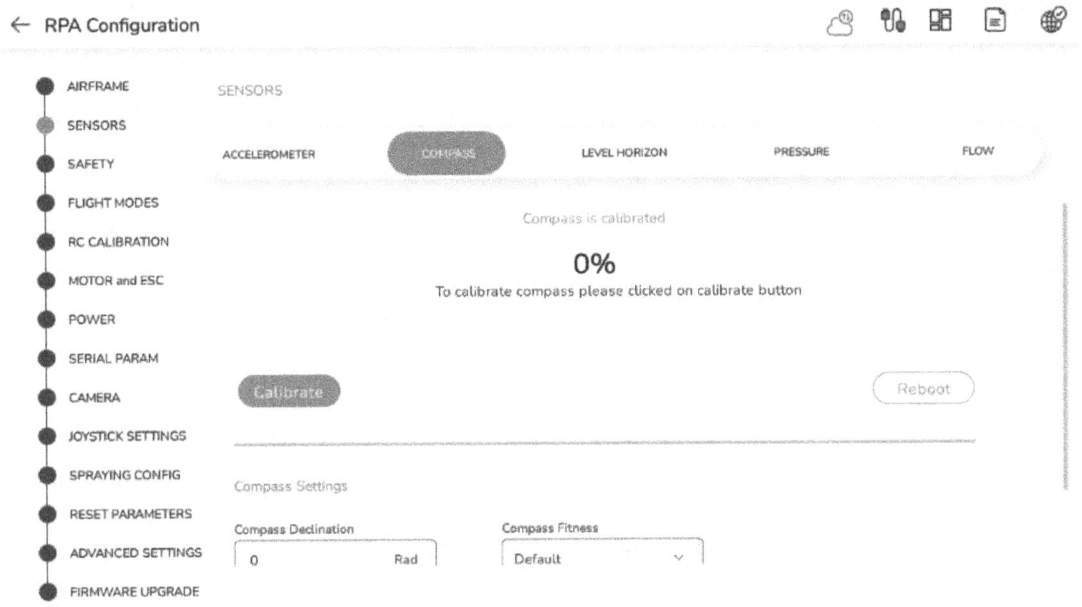

Figure 12.11 – The COMPASS window

5. Then, we have level horizon calibration. This calibrates the level of the flight controller. It helps adjust any minor misalignments when it's on a level surface to achieve the appropriate level in the ground station flight view. This form of calibration is highly recommended for optimal flight performance.

Place the drone on a level surface and click **LEVEL HORIZON**. The level horizon calibration process will automatically complete:

256　Developing a Custom Survey Drone

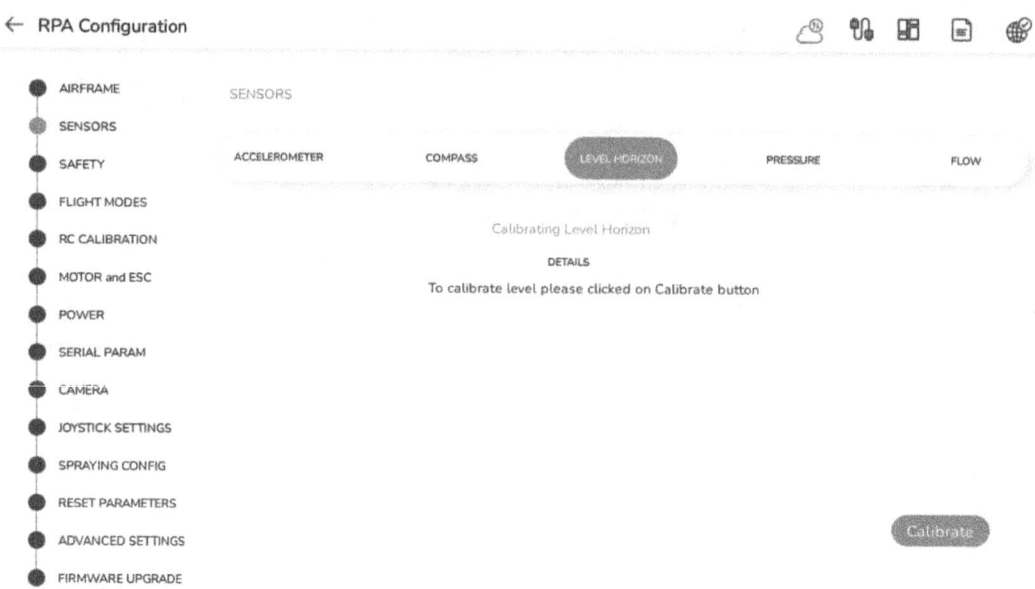

Figure 12.12 – The LEVEL HORIZON window

6. Next, we have pressure/barometer calibration. Calibrating the pressure sensor sets the drone's current altitude to 0 at the current pressure:

- Click on the **Calibrate** button; calibration will take place
- Once the calibration process has finished, a message stating **Barometer calibration complete** will appear on the screen

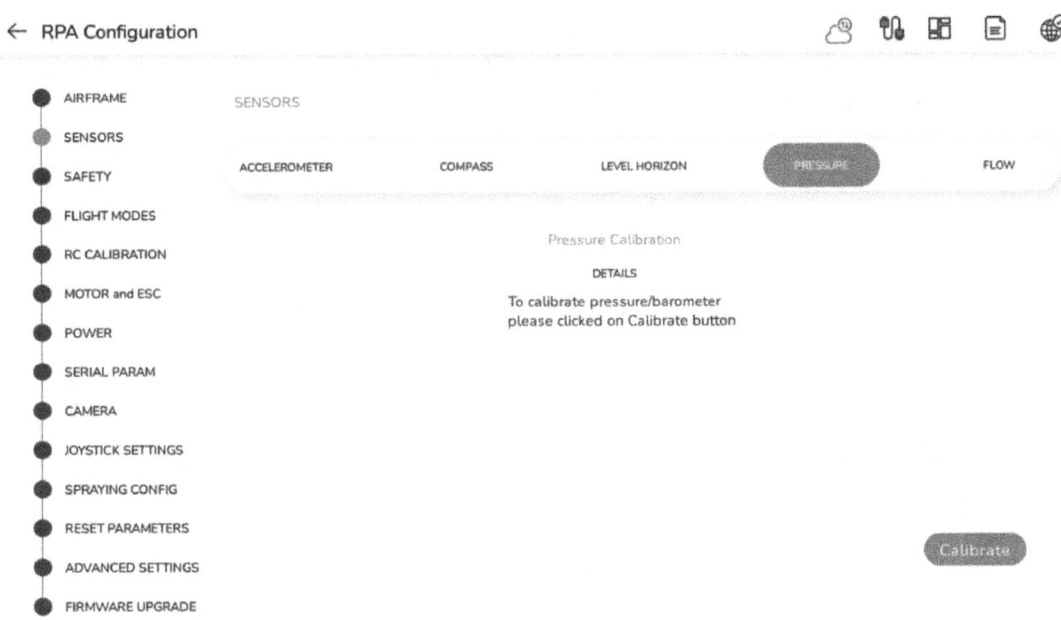

Figure 12.13 – The PRESSURE window

7. Next, we must set the safety parameters:

    - Battery failsafe trigger:

        + Low Action: RTL (Return to Launch)
        + Critical Action: RTL (Return to Launch)
        + Low Voltage Threshold: 21 V
        + Critical voltage threshold: 20 V

    - Failsafe trigger:

        + Ground Station Failsafe: Always enable RTL
        + Throttle Failsafe: Always RTL

    - PWM threshold:

        + Set PWM Threshold to 1000

    - RTL (Return to Launch)

Adjust these settings according to your requirements. However, for survey missions, the altitude should stay between 80 and 150m AGL. Hence, it is recommended to choose **Return At Current Altitude**. The rest of the settings can remain the same:

Figure 12.14 – The RTL window

Keep the other settings as is or go through the AeroGCS manual and adjust the settings according to your needs.

1. Next, we will assign the flight mode. Keep **Flight Mode Channel** set to **Channel 5** since this is the flight stack standard. Note that this isn't mandatory. Users can choose a different one as per their requirements.

Assign different flight modes to the channel so that it can be configured on RC/Herelink as well:

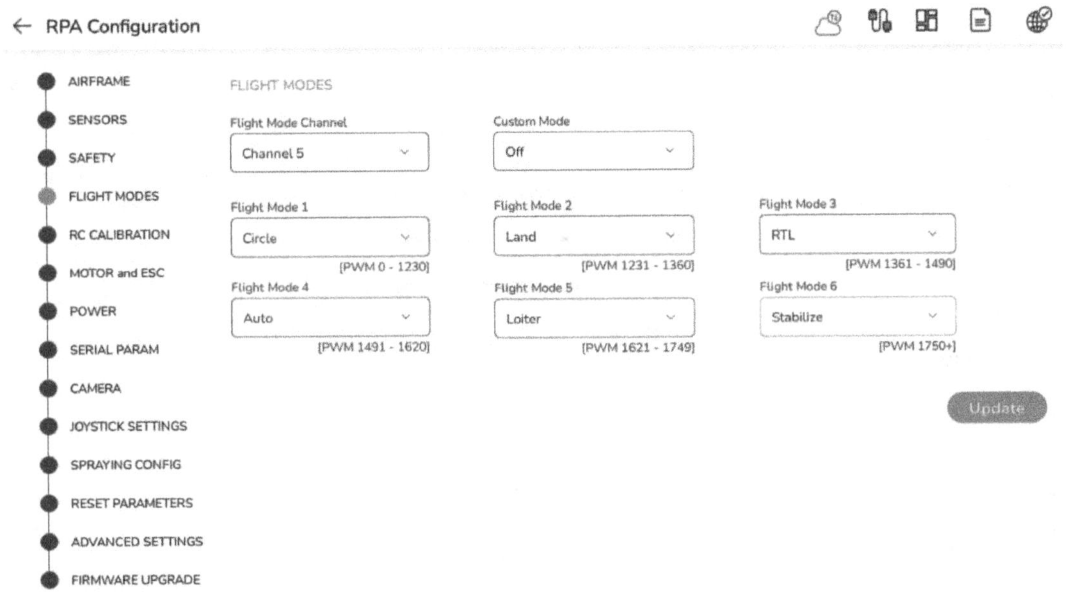

Figure 12.15 – The FLIGHT MODES window

2. Finally, we have RC calibration. For RC configuration (**RPA Configuration | RC Config**), RC transmitters allow the pilot to set the flight mode, control the vehicle's movement and orientation, and also turn on/off auxiliary functions (that is, raising and lowering the landing gear, and so on).

   RC calibration involves capturing each RC input channel's minimum, maximum, and "trim" values so that ArduPilot can correctly interpret the input.

   Move both sticks in the largest circle possible so that they reach their complete range of motion. Move the Channel 5 and 6 toggle switches through their range of positions.

   Your transmitter should cause the following control changes:

   - **Channel 1**: Low = roll left, high = roll right
   - **Channel 2**: Low = pitch forward, high = pitch back
   - **Channel 3**: Low = throttle down (off), high = throttle up
   - **Channel 4**: Low = yaw left, high = yaw right

Herelink also has a built-in RC calibration tool that can be used for RC calibration:

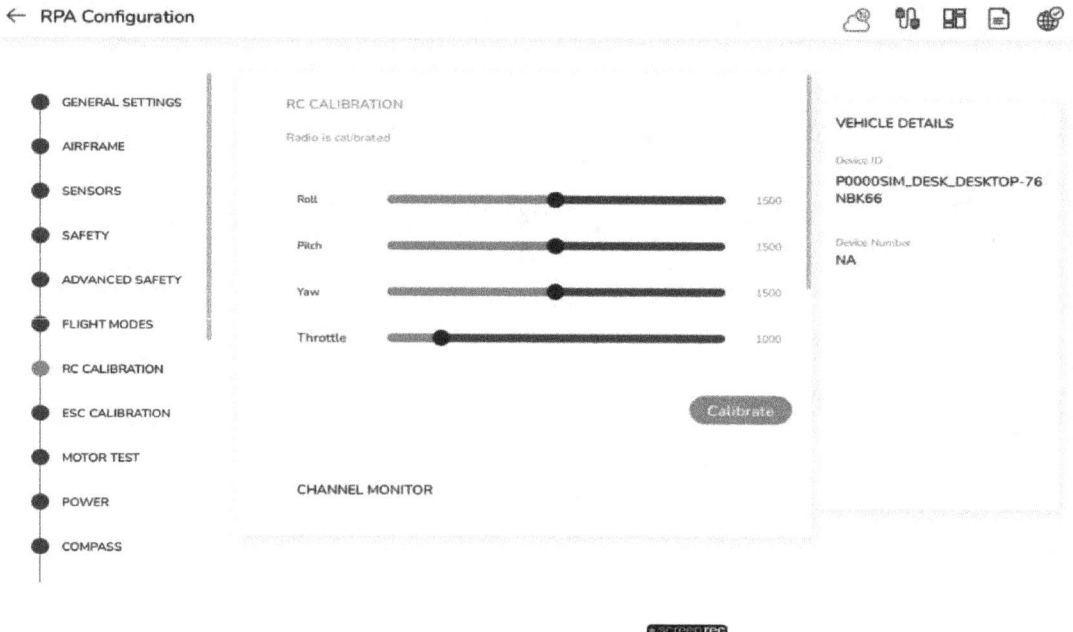

Figure 12.16 – The RC CALIBRATION window

Motor and ESC calibration

The **MOTOR and ESC** tab is used to test the motors and ESCs in different directions. Follow the instructions in the software to calibrate the ESCs and test the motors:

Drone configurations   261

Figure 12.17 – The MOTOR TEST window

ESC calibration

ESCs regulate motor speed (and direction) based on the PWM input value from the flight controller. The range of inputs to which an ESC will respond is configurable, and the default range can differ even between ESCs of the same model.

This calibration process updates all the ESCs with the maximum and minimum PWM input values that will be supplied by the flight controller. Subsequently, all the ESCs/motors will respond to the flight controller's input in the same way (across the whole input range):

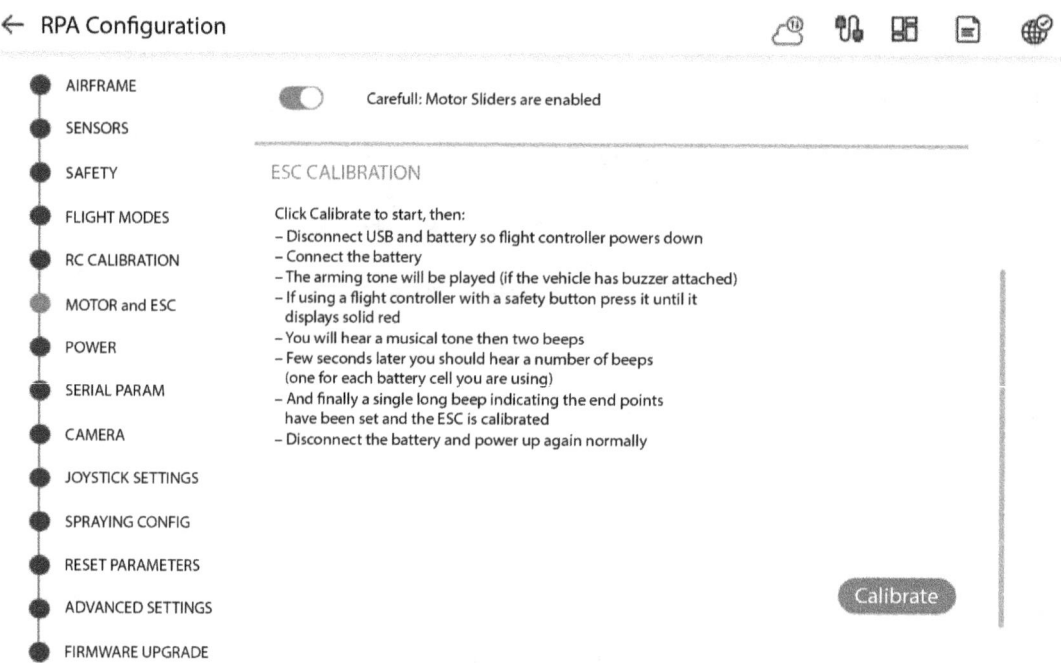

Figure 12.18 – The ESC CALIBRATION window

The steps for calibrating ESCs are as follows:

1. Click on the **Calibrate** button and proceed with the instructions that appear on the screen. Follow all the steps to calibrate all the ESCs.

2. Disconnect the USB and battery so that the flight controller powers down.

3. Connect the battery. The arming tone will be played if the vehicle has a buzzer attached.

4. Press the safety button (if the flight controller has one) until it turns a solid red color.

5. Listen out for two beeps that play a musical tone.

6. After a few seconds, listen out for several beeps (one for each battery cell you are using).

7. Listen out for a final single long beep. This indicates the endpoints and that the ESCs have been calibrated.

8. Disconnect the battery and power up again normally.

By following these instructions, we can successfully configure a drone. This guide was just an introduction to different settings. The skills you gain depend on practice and reading. The more we practice, the more we understand things. You will encounter many hurdles on your first try.

The preceding configuration can also be done with Herelink and its internally built QGC app. Please follow the Herelink guide to learn how to configure a drone with it.

Summary

In this chapter, we finished developing a survey drone. The important part of this process is selecting the correct components.

We have seen that a basic airframe can be chosen; however, we can design and improve the aesthetics of the drone as we want. We also studied drone configurations through AeroGCS. We used the Herelink system for telemetry and RC control, and its configuration can be learned through its datasheet. Then, we took a quick look at what we must do after assembling a drone. As mentioned previously, practice makes perfect; the more we study and practice, the more we'll be close to perfection. Getting into the habit of studying datasheets and integration manuals will be highly beneficial in this regard.

References

This book was written with the help and assistance of resources from authors of other books. I also give due credit to all the resources and images that helped to complete the book and allowed me to deliver it to you. The following are the references and resources used in each chapter.

Chapter 1, Getting Started with UAV and Drone Engineering

- *Introduction to Unmanned Aircraft Systems*: https://ftp.idu.ac.id/wp-content/uploads/ebook/tdg/MILITARY%20PLATFORM%20DESIGN/Unmanned%20Aircraft%20Systems.pdf
- *Applications of Unmanned Aerial Vehicles: A Review*: https://3ciencias.com/wp-content/uploads/2019/11/art-5_special-issue_3C-TECNO_november_2019-1.pdf
- *Different Types of Drones and Uses (2024 Full Guide)*: https://www.jouav.com/blog/drone-types.html
- *Drone Types: Multi-Rotor Vs Fixed-Wing Vs Single Rotor Vs Hybrid VTOL*: https://www.auav.com.au/articles/drone-types/
- *What are the parts of a Drone? Full list*: https://umilesgroup.com/en/what-are-the-parts-of-a-drone-full-list/
- *Anatomy of A Drone - What's inside a DJI Phantom Drone*: https://www.dronefly.com/the-anatomy-of-a-drone
- *Bicopter - The journey so far*: https://discuss.ardupilot.org/t/bicopter-the-journey-so-far/59328
- *From drones to geospatial analysis*: https://www.researchgate.net/figure/Tricopter-unmanned-aerial-vehicle_fig4_330555127
- *DJI F450 Quadcopter Review*: https://www.best-quadcopter.com/reviews/2020/10/dji-f450-quadcopter-review/
- *Hexacopter*: https://udayton.edu/engineering/research/centers/vision_lab/research/robotics_and_vision_guided_navigation/hexacopter.php

References

- *OCTA-QUAD X8 Coaxial Quad Design Notes for Arducopter*: https://www.rcgroups.com/forums/showthread.php?2952951-OCTA-QUAD-X8-Coaxial-Quad-Design-Notes-for-Arducopter

- *Hanseatic-S360 fixed wing drone*: https://www.routescene.com/resources/recommended-uav-drones-lidar/attachment/hanseatic-s360-fixed-wing-drone/

- *Development and Experimental Verification of a Hybrid Vertical Take-Off and Landing (VTOL) Unmanned Aerial Vehicle(UAV)*: https://www.researchgate.net/figure/Hybrid-VTOL-UAV-explanation_fig2_317933218

- *The application of the Vtol fixed-wing drone is the future*: https://www.dronefromchina.com/new/The-application-of-the-Vtol-fixed-wing-drone-is-the-future.html

- *Control Concept of a Tiltwing UAV During Low Speed Manoeuvring*: https://www.semanticscholar.org/paper/Control-Concept-of-a-Tiltwing-UAV-During-Low-Speed-Ostermann-Holsten/de6f95fa77be85083ca1fe02a94ffc44f7de58e4

- *Airframes*: https://www.google.com/url?sa=i&url=https%3A%2F%2Fwww.alibaba.com%2Fpla%2FJMRRC-6-rotor-Multi-rotor-Hexacopter-Frame-960mm_60280242487.html%3Fmark%3Dgoogle_shopping%26biz%3Dpla%26searchText%3Daircraft%26product_id%3D60280242487%26language%3Den&psig=AOvVaw3ATLoSpH7XCXC0ycDhCiJ-&ust=1710349518571000&source=images&cd=vfe&opi=89978449&ved=0CBMQjRxqFwoTCID38M2a74QDFQAAAAAdAAAAABAI

- *LiPo battery*: https://robokits.co.in/batteries-chargers/genx-power-premium-lipo-battery/genxpower-29.6v-lipo-batteries/genx-29.6v-8s-16000mah-25c-50c-premium-lipo-battery-with-as150-connector

- *PDB*: https://gadgetsdeal.in/shop/pdb/apd-pdb500-x-12s-52v-500a-board-in-india/

- *Buck and boost converters*: https://electronicshub.pk/product/xl6009-dc-to-dc-boost-converter-module-in-pakistan/

- *Transmitter*: https://robu.in/product/radiolink-at10-2-4ghz-12ch-rc-drone-transmitter/

- *Handheld GCS*: https://www.aero-sentinel.com/military-drones/drone-gcs/

- *Radio modem*: https://www.dronefromchina.com/product/RFD900X-900MHz-Ultra-Long-Range-up-to-40KM-Radio-Data-Modem.html

- GPS: https://baskaerospace.com.au/shop/gps/here3/
- Compass/magnetometer: https://drotek.gitbook.io/additional-devices/sensors/geomagnetic-sensor/rm3100
- Flight controller: https://robu.in/product-category/drone-parts/flight-controller-accessories/
- Payload: https://forum.iktva.sa/drone-camera-gimbal-30x-uav-gimbal-zoom-hd-starlight-ww-18979184
- https://www.jmrdrone.com/agricultural-landing-gear/
- https://books.google.ae/books?id=2M0hEAAAQBAJ&printsec=copyright&redir_esc=y#v=onepage&q&f=false

Chapter 2, Understanding the Dynamics of 3D Space

- *Working Principle and Components of Drone*: https://cfdflowengineering.com/working-principle-and-components-of-drone/
- *Forces active in drone flight*: https://www.dronesindevelopment.com/post/forces-active-in-drone-flight
- *Motion basics: How to define roll, pitch, and yaw for linear systems*: https://www.linearmotiontips.com/motion-basics-how-to-define-roll-pitch-and-yaw-for-linear-systems/
- *Roll, Pitch & Yaw Explained for Beginners*: https://www.dronesvilla.com/roll-pitch-yaw/
- *Drone dictionary — essential terms every pilot should know*: https://ageagle.com/blog/drone-dictionary-essential-terms-every-pilot-should-know/#:~:text=Roll%3A%20an%20aircraft's%20rotation%20about,travel%20determined%20by%20the%20FAA
- *An A-Z Glossary of Drone Terminology*: https://www.hovrtek.com/drone-operations/drone-terminology/
- *Drone terminology reference guide*: https://www.heliguy.com/blogs/posts/drone-terminology-reference-guide
- *Demystifying Drone Dynamics!*: https://towardsdatascience.com/demystifying-drone-dynamics-ee98b1ba882f
- *Quadcopter Dynamics*: https://catsr.vse.gmu.edu/SYST460/QuadcopterDynamics.pdf

- *Basics of Unmanned Aerial Vehicles: Time to start working on Drone Technology*: https://books.google.ae/books?id=2M0hEAAAQBAJ&printsec=copyright&redir_esc=y#v=onepage&q&f=false

- *Working Principle and Components of Drone*: https://cfdflowengineering.com/working-principle-and-components-of-drone/

- *Flight Control*: https://developer.dji.com/mobilesdk/documentation/introduction/flightController_concepts.html

- *21 Basic Mechanics University of Pennsylvania Coursera*: https://www.youtube.com/watch?v=ZEN4X185Z8M&list=PLblGgzWkqSqM7IWsgjDetdzZDS1NbkTnd&index=4&pp=iAQB

Chapter 3, Learning and Applying Basic Command and Control Interfaces

- *Ground Control Software*: https://www.unmannedsystemstechnology.com/expo/ground-control-software/

- *UAV Ground Control Station: The Nerve Center of Drone Operations*: https://defensebridge.com/article/uav-ground-control-station-the-nerve-center-of-drone-operations.html

- *Choosing a Ground Station*: https://ardupilot.org/plane/docs/common-choosing-a-ground-station.html

- *LaunchPad GCS*: https://www.urbanmatrix.co.in/umt-launchpad

- *QGroundControl: Intuitive and Powerful Ground Control Station for the MAVLink protocol*: http://qgroundcontrol.com/

- *AeroGCS*: https://aeromegh.com/aerogcs-kea/

- *Take on better-paid and more complex projects for drone surveys*: https://www.sphengineering.com/flight-planning/ugcs

- *MAVLink Developer Guide*: https://mavlink.io/en/

- *MAVLink Basics*: https://ardupilot.org/dev/docs/mavlink-basics.html

- *What are Communication Protocols & Their Working*: https://www.elprocus.com/communication-protocols/

- *Top 6 IoT Communication Protocols*: https://wisilica.com/company/top-6-iot-communication-protocols/

- *Communication protocols in IoT: Introduction and Guide*: https://www.cdebyte.com/news/476

- *Top 12 most commonly used IoT protocols and standards*: https://www.techtarget.com/iotagenda/tip/Top-12-most-commonly-used-IoT-protocols-and-standards
- *RC*: https://www.frsky-rc.com/
- *Drone components*: https://www.flysky-cn.com/
- *Drone Controllers: A Look at How They Work, Important Terminology, and Why They're Unique in the RC Aircraft World*: https://uavcoach.com/drone-controller/
- *Drone Transmitter and Receiver Guide*: https://robocraze.com/blogs/post/drone-transmitter-and-receiver-guide
- *Radio Control Systems*: https://docs.px4.io/v1.9.0/en/getting_started/rc_transmitter_receiver.html
- *FPV Protocols Explained (CRSF, SBUS, DSHOT, ACCST, PPM, PWM and more)*: https://oscarliang.com/rc-protocols/#:~:text=Latency-,What%20are%20RC%20Protocols%3F,in%20another%20language%20(output)
- https://blog.banggood.com/top-7-rc-protocols-that-you-should-know-about-to-sound-quad-smart-33056.html

Chapter 4, Knowing UAV Systems, Sub-Systems, and Components

- MN5006 Antigravity Type 4-6S UAV Motor KV300: https://store.tmotor.com/product/mn5006-kv300-motor-antigravity-type.html
- *Electronic Speed Controller ESC*: https://www.electronicclinic.com/electronic-speed-controller-esc/
- *Drone Design - Calculations and Assumptions*: https://www.tytorobotics.com/blogs/articles/the-drone-design-loop-for-brushless-motors-and-propellers
- *Design and Implementation of a Real Time Wireless Quadcopter for Rescue Operations*: https://www.researchgate.net/publication/324391001_Design_and_Implementation_of_a_Real_Time_Wireless_Quadcopter_for_Rescue_Operations
- *Average Weights of Common Types of Drones*: https://www.droneblog.com/average-weights-of-common-types-of-drones/
- *How Much Does Drone Weight? Importance Of Correct Weight*: https://www.propelrc.com/how-much-does-drone-weight/

- *Drone Motor Calculator*: https://www.omnicalculator.com/other/drone-motor
- *How to select drone motors*: https://www.engineersgarage.com/how-to-select-drone-motors/
- *A Guide to Understanding LiPo Batteries*: https://www.rogershobbycenter.com/lipoguide
- *The Terminology, Formulae and Physical Assembly of Lipo Batteries - or - Wtf does "C" mean?*: https://electric-skateboard.builders/t/the-terminology-formulae-and-physical-assembly-of-lipo-batteries-or-wtf-does-c-mean/33162
- *Material Selection for Drones (UAV), Degradation and Thermal Stress Calculations*: https://www.studocu.com/row/document/jamaa%D8%A9-alemarat-alaarby%D8%A9-almthd%D8%A9/engineering-material/material-selection-for-drones-uav-degradation-and-thermal-stress-calculations/10510339
- *How To Choose The Best Drone Frame - Drone Frames Guide*: https://www.jinjiuyi.net/news/how-to-choose-the-best-drone-frame.html
- *How to Choose ESC for Quadcopter*: https://robocraze.com/blogs/post/how-to-choose-esc-for-quadcopter
- *How to Select your Drone Flight Controller – A Comparative Selection Guide*: https://circuitdigest.com/article/how-to-select-your-drone-flight-controller-a-comparative-selection-guide
- *Definitive Guide on How to Choose the Flight Controller for Your Racing Drone FPV*: https://www.drone24hours.com/blog/come-scegliere-flight-controller-per-quadcopter/?lang=en
- *UAV Ground Control Stations*: https://www.unmannedsystemstechnology.com/expo/ground-control-stations-gcs/
- *GNSS Basics for the Drone Mapper*: https://microaerialprojects.com/wp-content/uploads/2018/10/GNSS-FOR-THE-DRONE-MAPPER-9-30-2018-1.pdf

Chapter 5, Sensors and IMUs with Their Application in Drones

- *Where are MEMS Gyroscopes Used?*: https://www.etransaxle.com/where-are-mems-gyroscopes-used
- https://www.google.com/url?sa=i&url=https%3A%2F%2Fwww.domainesia.com%2Fberita%2Fapa-itu-accelerometer-fungsi%2F&psig=AOvVaw3hqf

- vXEc79K4LRTOQy00SL&ust=1697825104753000&source=images&cd=vfe&opi=89978449&ved=0CBEQjRxqFwoTCJjr4sPZgoIDFQAAAAAdAAAAABAE

- *Gyroscope*: https://www.google.com/url?sa=i&url=https%3A%2F%2Fen.wikipedia.org%2Fwiki%2FGyroscope&psig=AOvVaw3aY9XdcNma8msAEXRz8ypL&ust=1697825271259000&source=images&cd=vfe&opi=89978449&ved=0CBEQjRxqFwoTCNCWk5PagoIDFQAAAAAdAAAAABAE

- *Calibration of a Magnetometer with Raspberry Pi*: https://makersportal.com/blog/calibration-of-a-magnetometer-with-raspberry-pi

- *What is a Barometer?*: https://instrumentationtools.com/what-is-a-barometer/

- *How It Works: PPK vs RTK Drone Surveying*: https://www.propelleraero.com/blog/how-it-works-ppk-vs-rtk-drone-surveying/

- *Inertial Measurement Unit (IMU) – An Introduction*: https://www.advancednavigation.com/tech-articles/inertial-measurement-unit-imu-an-introduction/

- *MEMS accelerometer*: https://www.electricity-magnetism.org/mems-accelerometer/

- *Barometers*: https://docs.px4.io/main/en/sensor/barometers.html

- *Magnetometers And Navigation*: https://www.geographyrealm.com/magnetometers-and-navigation/

- *UAV Navigation In Depth: Magnetometer, Why Is It Critical For UAV Navigation?*: https://www.uavnavigation.com/company/blog/uav-navigation-depth-magnetometer

Chapter 6, Introduction to Drone Firmware and Flight Stacks

- Smith, John (2020). *Understanding Firmware: The Backbone of Electronic Devices.* New York: TechPress

- Patel, Rajesh (2019). *Exploring Firmware: An Overview of Embedded Software Development.* Journal of Embedded Systems Engineering, 5(2), 78-92

- Anderson, Mark (2017). *Introduction to Firmware Programming.* San Francisco: O'Reilly Media

- Thompson, Paul (2018). *Types of Firmware: A Comprehensive Overview.* Journal of Electronic Engineering, 12(1), 23-36

- Gupta, Ankit (2016). *Firmware Engineering: Concepts and Practices.* London: Springer

- Li, Xiaowei (2015). *The Evolution of Firmware: From Basic Input/Output Systems to Modern Embedded Software.* IEEE Computer Society Magazine, 29(4), 32-45

- Johnson, Michael (2020). *Embedded Systems Programming: A Practical Guide.* Cambridge: Cambridge University Press

- Brown, David, and Chen, Wei (2019). *Firmware Security: Challenges and Best Practices.* Sebastopol: O'Reilly Media

- Williams, Emily (2018). *Embedded Systems Design: Principles and Applications.* Boston: Academic Press

- Patel, Rajesh (2020). *Advancements in Drone Flight Stacks.* IEEE Robotics and Automation Letters, 5(3), 567-580

- ArduPilot (n.d.). *About ArduPilot.* Retrieved from `https://ardupilot.org/about`

- ArduPilot Development Team (2021). *ArduPilot: An Open Source UAV Platform.* Journal of Unmanned Aerial Systems, 8(2), 45-58

- ArduPilot documentation (n.d.). *ArduPilot Architecture Overview.* Retrieved from `https://ardupilot.org/dev/docs/learn-about-the-architecture-of-ardupilot.html`

- Anderson, Mark (2019). *Understanding the Architecture of ArduPilot: A Comprehensive Guide.* Journal of Robotics and Automation, 10(1), 89-102

- PX4 Autopilot (n.d.). *Introduction to PX4.* Retrieved from `https://docs.px4.io/master/en/`

- Anderson, Mark (2022). *PX4 Autopilot: An Open Source Flight Control Software for Drones.* Sebastopol: O'Reilly Media

- PX4 Developer Guide (n.d.). *PX4 Architecture Overview.* Retrieved from `https://dev.px4.io/master/en/`

- Brown, David, and Chen, Wei (2020). *An In-Depth Look into PX4 Architecture.* Journal of Unmanned Aerial Vehicles, 6(4), 210-225

- PX4 Autopilot Development Team (n.d.). *PX4 Developer Guide: Control Architecture.* Retrieved from `https://dev.px4.io/v1.11/en/concept/architecture.html#control-architecture`

- Anderson, Mark (2019). *Flight Control Systems for Drones: Algorithms and Implementation.* San Francisco: O'Reilly Media

Chapter 7, Introduction to Ground Control Station Software

- Smith, John (2020). *Introduction to Ground Control Stations: Principles and Applications.* New York: TechBooks

- Johnson, Michael (2018). *Understanding Ground Control Stations for UAV Operations.* Journal of Unmanned Aerial Systems, 6(3), 45-60

- *Ground Control Stations for Unmanned Aerial Vehicles, Drones and Remotely Piloted Aircraft Systems*: https://www.unmannedsystemstechnology.com/expo/ground-control-stations-gcs/#:~:text=What%20are%20Ground%20Control%20Stations,direct%20control%20of%20the%20UAV

- Anderson, Mark (2019). *Drone Control Software and Hardware: A Comprehensive Guide.* San Francisco: O'Reilly Media.

- Williams, Emily (2021). *Comparative Analysis of GCS Software and Hardware Options.* International Journal of Aerial Robotics, 18(2), 167-182

- *Choosing a Ground Station*: https://ardupilot.org/plane/docs/common-choosing-a-ground-station.html

- *Ground Control Stations*: https://www.uxvtechnologies.com/ground-control-stations/

- *Ground Control Station Background*: https://uavrt.nau.edu/index.php/docs/control/gcsbackground/

- Mission Planner Development Team (n.d.). *Mission Planner User Guide.* Retrieved from https://ardupilot.org/planner/docs/mission-planner-overview.html

- Brown, David (2018). *Mission Planner: A Comprehensive Overview.* Journal of UAV Applications, 10(2), 112-128

- APM Planner Development Team (n.d.). *APM Planner User Manual.* Retrieved from https://ardupilot.org/planner2/docs/intro.html

- Simpson, John (2016). *APM Planner: Features and Functionality.* Journal of Unmanned Systems, 8(3), 210-225

- QGroundControl Development Team (n.d.). QGroundControl User Guide. Retrieved from https://docs.qgroundcontrol.com/en/

- Gupta, Ankit (2020). *QGroundControl: An Open-Source GCS Platform.* Journal of Unmanned Vehicle Systems, 15(4), 145-160

- UGCS Development Team (n.d.). *UGCS User Manual.* Retrieved from https://www.ugcs.com/

- Chen, Wei (2017). *UGCS: Features and Capabilities.* Journal of Remote Sensing and GIS, 20(1), 89-104

- AeroGCS: https://aeromegh.com/aerogcs-kea/

- AeroGCS GREEN: https://aeromegh.com/aerogcs-green/

- Aero GCS GREEN Installation link: `https://play.google.com/store/apps/detailsid=com.aerogcs.aerogcsenterprise.hellodrone&hl=en_US&pli=1`
- *AeroGCS KEA 2.3*: `https://aerogcs-docs.aeromegh.com/`

Chapter 8, Understanding Flight Modes and Mission Planning

- *Mastering your Drone's Flight Modes: A Beginner's Guide*: `https://www.womenwhodrone.co/single-post/mastering-your-drone-s-flight-modes-a-beginner-s-guide`
- *Drone Flight Modes*: `https://www.dummies.com/article/technology/electronics/drones/drone-flight-modes-142437/`
- *Flight Modes (Multicopter)*: `https://docs.px4.io/main/en/flight_modes_mc/`
- *Flight Modes (Developers)*: `https://docs.px4.io/main/en/concept/flight_modes.html`
- *Flight Modes*: `https://ardupilot.org/copter/docs/flight-modes.html`
- *DRONE flight manual*: `https://escholarship.org/content/qt2zv0z6zm/qt2zv0z6zm_noSplash_51fd3dcbdb109ea18d5efccd5aa94a90.pdf?t=qv9ey7`
- *The Droner's Manual*: `https://cdn11.bigcommerce.com/s-m5qljysoqy/content/look-inside/UAS-DRONE.pdf`
- *Position Mode (Multicopter)*: `https://docs.px4.io/main/en/flight_modes_mc/position.html`
- *Altitude Mode (Multicopter)*: `https://docs.px4.io/main/en/flight_modes_mc/altitude.html`
- *Manual/Stabilized Mode (Multicopter)*: `https://docs.px4.io/main/en/flight_modes_mc/manual_stabilized.html`
- *Takeoff Mode (Multicopter)*: `https://docs.px4.io/main/en/flight_modes_mc/takeoff.html`
- *Land Mode (Multicopter)*: `https://docs.px4.io/main/en/flight_modes_mc/land.html`
- *Hold Mode (Multicopter)*: `https://docs.px4.io/main/en/flight_modes_mc/hold.html`
- *Mission Mode (Multicopter)*: `https://docs.px4.io/main/en/flight_modes_mc/mission.html`

- *Return Mode (Multicopter)*: https://docs.px4.io/main/en/flight_modes_mc/return.html
- *RC Transmitter Flight Mode Configuration*: https://ardupilot.org/copter/docs/common-rc-transmitter-flight-mode-configuration.html
- *Flight Mode Configuration*: https://docs.px4.io/main/en/config/flight_mode.html
- *Assigning Flight Modes*: https://aerogcs-docs.aeromegh.com/5-rpa-configuration/6.4-assigning-flight-modes
- *Mission Success: Mission Planning and Execution Software in Drone Operations*: https://www.linkedin.com/pulse/mission-success-planning-execution-software-drone-operations
- *What are the steps to plan and execute a successful drone mission?*: https://www.linkedin.com/advice/0/what-steps-plan-execute-successful-drone-mission-skills-drones
- *Planning a Mission with Waypoints and Events*: https://ardupilot.org/copter/docs/common-planning-a-mission-with-waypoints-and-events.html
- *Waypoint Planning*: https://aerogcs-docs.aeromegh.com/6-project-and-flights/7.1-waypoint-planning

Chapter 9, Drone Assembly, Configuration, and Tuning

- *Drone Part List*: https://robocraze.com/blogs/post/drone-part-list
- *What are the parts of a Drone? Full list*: https://umilesgroup.com/en/what-are-the-parts-of-a-drone-full-list/
- *How to Make a Quadcopter Drone and Components List*: https://www.instructables.com/How-to-Make-a-Quadcopter-Drone-and-Components-List/
- *FPV Tools and Materials for Building Drones and Fixed Wings*: https://oscarliang.com/fpv-tools/
- *What Tools are Needed to Build a Drone?*: https://www.heamar.co.uk/blog/68_what-tools-are-needed-to-build-a-drone
- *Tarot Frames*: https://www.3dxr.co.uk/multirotor-c3/multirotor-frames-c97/tarot-m36
- *Build a High Performance FPV Camera Quadcopter*: https://www.instructables.com/Build-a-High-Performance-FPV-Camera-Quadcopter/

- *Assembling, Calibration and Tuning of Quadcopter*: https://nesac.gov.in/assets/resources/2020/12/Assembling-Calibration-and-Tuning-of-Quadcopter.pdf
- *Mounting ESCs and arms*: https://nxp.gitbook.io/hovergames/archive/s500-drone-frame/mounting-escs-and-arms
- *Electronic Speed Controller (ESC): Everything You Need to Know*: https://www.jouav.com/blog/electronic-speed-controller-esc.html
- *Safety Precaution for Handling of Carbon Fiber*: http://www.3gcarbon.com/news/shownews.php?id=42
- *Build Your Own FPV Drone: A Beginner's Guide*: https://blogs.cuit.columbia.edu/mep14/fpv-drone/build-your-own-fpv-drone-a-beginners-guide/
- *Loading Firmware onto boards without existing ArduPilot firmware*: https://ardupilot.org/copter/docs/common-loading-firmware-onto-chibios-only-boards.html
- *Loading Firmware*: https://ardupilot.org/copter/docs/common-loading-firmware-onto-pixhawk.html
- *Loading Firmware*: https://docs.px4.io/main/en/config/firmware.html
- *Connecting a Drone*: https://aerogcs-docs.aeromegh.com/5-connecting-a-drone
- *Mandatory Hardware Configuration*: https://ardupilot.org/copter/docs/configuring-hardware.html
- *Safety Configuration (Failsafes)*: https://docs.px4.io/main/en/config/safety.html
- *Manual Firmware Upgrade*: https://aerogcs-docs.aeromegh.com/5-rpa-configuration/6.14-firmware-upgrade/6.14.2-manual-firmware-upgrade
- *Calibration of Sensors*: https://aerogcs-docs.aeromegh.com/5-rpa-configuration/6.2-calibration-of-sensors
- *Safety Parameters*: https://aerogcs-docs.aeromegh.com/5-rpa-configuration/6.3-safety-parameters
- *First Flight Guidelines*: https://docs.px4.io/main/en/flying/first_flight_guidelines.html
- *First Flight Tutorial For The VOXL 2-Based PX4 Autonomy Developer Kit*: https://px4.io/first-flight-tutorial-for-the-voxl-2-based-px4-autonomy-developer-kit/

Chapter 10, Flight Log Analysis and PIDs

- *Open Source Flight Log Analysis for Drone Diagnostics*: https://www.unmannedsystemstechnology.com/2023/11/open-source-flight-log-analysis-for-drone-diagnostics/#:~:text=Tilak.io%2C%20a%20specialist%20developer,3D%20to%20provide%20actionable%20insights
- Airdata UAV, Get Started: https://app.airdata.com/main?a=upload
- Drone flight log: https://dronedeck.eu/drone-flight-log/
- *Flight Log Analysis*: https://docs.px4.io/main/en/log/flight_log_analysis.html
- *Flight Log Analysis – Import, analysis, and plotting of telemetry logs*: https://www.mathworks.com/help/uav/flight-log-analysis.html
- *DroneePlotter - Drone Flight Log Analysis Tool Works on Web Browser*: https://diydrones.com/profiles/blogs/droneeplotter-drone-flight-log-analysis-tool-works-on-web-browser
- *Drone Operations Compliance & Fleet Management Made Easy*: https://www.dronelogbook.com/hp/1/index.html
- Flight Review: https://review.px4.io/
- *Log Analysis using Flight Review*: https://docs.px4.io/main/en/log/flight_review.html
- *Downloading and Analyzing Data Logs in Mission Planner*: https://ardupilot.org/copter/docs/common-downloading-and-analyzing-data-logs-in-mission-planner.html
- *Analyze View*: https://docs.qgroundcontrol.com/Stable_V4.3/en/qgc-user-guide/analyze_view/index.html#:~:text=The%20Analyze%20View%20is%20accessed,clear%20logs%20on%20the%20vehicle
- *UAV LogViewer*: https://ardupilot.org/dev/docs/common-uavlogviewer.html
- *Vibration Analysis of a Quadcopter Carrying a Payload*: https://www.researchgate.net/publication/356682905_Vibration_Analysis_of_a_Quadcopter_Carrying_a_Payload
- *Log Analysis using Flight Review*: https://docs.px4.io/main/en/log/flight_review.html
- *What is a PID Controller?*: https://www.omega.com/en-us/resources/pid-controllers#:~:text=A%20PID%20(Proportional%20%E2%80%93%20

Integral%20%E2%80%93,variables%20in%20industrial%20control%20
  systems

- *Introduction: PID Controller Design*: https://ctms.engin.umich.edu/CTMS/index.
  php?example=Introduction&section=ControlPID

- *Multicopter PID Tuning Guide (Manual/Advanced)*: https://docs.px4.io/main/en/
  config_mc/pid_tuning_guide_multicopter.html

- *PID Flight Controller for Quadcopter and its Simulation*: https://ijirt.org/master/
  publishedpaper/IJIRT160429_PAPER.pdf

Chapter 11, Application-Based Drone Development

- *Drone Surveying: Best Survey Drones, How to Do Drone Surveying, and More [New for 2024]*: https://uavcoach.com/drone-surveying/

- *Surveying & GIS*: https://wingtra.com/drone-mapping-applications/
  surveying-gis/

- *Drone Surveying: Why it's Important and How it Works*: https://www.propelleraero.
  com/blog/drone-surveying-why-its-important-and-how-it-works/

- *Five Benefits of Using Drone Surveys*: https://www.landform-surveys.co.uk/
  news/five-benefits-of-using-drone-surveys/

- *Sensor payloads for small unmanned aerial vehicles*: https://www.osti.gov/servlets/
  purl/1498522

- *Sensors and sensitivity: drone sensor payloads used at mine sites*: https://www.
  measureaustralia.com.au/news/c-mines/sensors-and-sensitivity-
  drone-sensor-payloads-used-at-mine-sites

- *PPK Or RTK – Which is best?*: https://coptrz.com/blog/ppk-or-rtk-which-
  is-best/#:~:text=Real%20Time%20Kinematic%20(RTK)%20is,it%20is%20
  collected%20and%20uploaded

- *AeroGCS KEA 2.3*: https://aerogcs-docs.aeromegh.com/

- *Image stitching*: https://en.wikipedia.org/wiki/Image_stitching

- *Drone Photogrammetry: How Drone Photos Turn into 3D Surveys*: https://www.
  propelleraero.com/blog/drone-photogrammetry-how-drone-photos-
  turn-into-3d-surveys/

- *What deliverables can be achieved from using a drone for mapping?*: https://www.heliguy.
  com/blogs/knowledge-base/what-deliverables-can-be-achieved-
  from-using-a-drone-for-mapping#:~:text=There%20are%20several%20
  deliverables%20that,areas%2C%20and%20create%20accurate%20maps

- *The Benefits of Drones in Agribusiness*: https://www.travelers.com/resources/business-industries/agribusiness/benefits-of-drones-in-agribusiness

- *The Benefits of Drones in Agriculture*: https://www.xagaustralia.com.au/agriculture-drone-benefits

- *The benefits and drawbacks of using drones for last-mile delivery*: https://www.linkedin.com/pulse/benefits-drawbacks-using-drones-last-mile

- *Drone Delivery: What it is and what it means for retailers*: https://www.insiderintelligence.com/insights/drone-delivery-services/

- *10 Major Pros & Cons of Unmanned Aerial Vehicle(UAV) Drones*: https://www.equinoxsdrones.com/10-major-pros-cons-of-unmanned-aerial-vehicleuav-drones/

Chapter 12, Development of Custom Survey Drones

- *The Benefits of Using Drones for Surveying and Mapping*: https://www.linkedin.com/pulse/benefits-using-drones-surveying-mapping-karim-cosslett

- *Mapping & Surveying*: https://www.jouav.com/industry/aerial-mapping-surveying

- *How can you develop custom payloads for specialized drones?*: https://www.linkedin.com/advice/0/how-can-you-develop-custom-payloads-specialized-drones-3cewc

- *What Motors Are Used in Drones & How to Choose It?*: https://www.t-drones.com/blog/what-motors-are-used-in-drones.html

- *How to select drone motors*: https://www.engineersgarage.com/how-to-select-drone-motors/

- *Selecting Quadcopter Motor: A Detailed Guide 2021*: https://robu.in/selecting-quadcopter-motor-a-detailed-guide-2021/

- *Electronic Speed Controller (ESC): Everything You Need to Know*: https://www.jouav.com/blog/electronic-speed-controller-esc.html#:~:text=In%20general%2C%20the%20ESC%20should,are%20equipped%20with%20temperature%20sensors

- *How to Choose ESC for Quadcopter*: https://robocraze.com/blogs/post/how-to-choose-esc-for-quadcopter

- *How to Choose an Electronics Speed Controller for Your Quadcopter*: https://robu.in/how-to-choose-esc-for-your-quadcopter/

- *How to Choose a Drone Battery*: https://www.unmannedsystemstechnology.com/feature/how-to-choose-a-drone-battery/
- *How to Choose Lithium Polymer Battery for your RC Drone*: https://robu.in/how-choose-lithium-polymer-battery-for-drone/
- *How to choose a drone battery?*: https://www.linkedin.com/pulse/how-choose-drone-battery-t-drones
- Tarot Multirotor Frames: https://www.3dxr.co.uk/multirotor-c3/multirotor-frames-c97/tarot-m36
- T-MOTOR: https://uav-en.tmotor.com/
- *Choosing the Right Propellers for Your Drone*: https://robocraze.com/blogs/post/choosing-the-right-propellers-for-your-drone
- *Definitive Guide on How to Choose a Power Distribution Board*: https://www.drone24hours.com/blog/guida-allacquisto-di-pdb/?lang=en#:~:text=The%20main%20aspect%20of%20safety,the%20maximum%20total%20current%20draw
- *5 Flight controller tips*: https://www.linkedin.com/advice/1/what-most-effective-way-select-program-flight-controller#:~:text=Make%20sure%20the%20flight%20controller,performance%2C%20and%20customization%20you%20need
- *How to choose a flight controller for drones*: https://www.flyrobo.in/blog/flight-controller
- *How to Select your Drone Flight Controller – A Comparative Selection Guide*: https://circuitdigest.com/article/how-to-select-your-drone-flight-controller-a-comparative-selection-guide
- *What Makes a Good Camera for Drone Surveys and Inspections*: https://www.propelleraero.com/blog/what-makes-a-good-camera-for-drone-surveys-and-inspections/
- *How To Pick The Best Camera For Drone Photogrammetry*: https://www.heliguy.com/blogs/posts/how-to-pick-the-best-camera-for-drone-photogrammetry
- *Herelink Overview*: https://docs.cubepilot.org/user-guides/herelink/herelink-overview
- Skydroid H16 Pro HD Video Transmission System: https://www.worldronemarket.com/product/skydroid-h16-pro-remote-control/
- *CubePilot Ecosystem Autopilot Wiring Diagram*: https://docs.cubepilot.org/user-guides/cubepilot-ecosystem/cubepilot-ecosystem-autopilot-wiring-diagram

- *RC Calibration*: https://aerogcs-docs.aeromegh.com/5-rpa-configuration/6.5-rc-calibration
- *Camera and Camera Config*: https://aerogcs-docs.aeromegh.com/5-rpa-configuration/6.9-camera-and-camera-config

Index

Symbols

3D models 227
3D Robotics (3DR) 102
12 motors (in Hexa)/deca-hexa copter 11

A

above-ground level (AGL) 87
above mean sea level (AMSL) 87
acceleration power spectral
    density (PSD) 210
accelerometer 82, 83
actuator controls FFT graph 208, 209
actuator estimate graph 207
Advanced Power Drives (APD) 245
aerial delivery drones 230
    features 230
aerial survey mission
    payloads used 219, 220
    planning 222-224
aerial surveys 218
    advantages 218
    final deliverables 225-227
    role, of drones 219

Aero Ground Control Station
    (AeroGCS) 185
    download link 127
    home menu 129-133
    main dashboard 127, 128
    overview 127
AeroMegh 128
Agisoft Metashape 225
    URL 225
agricultural spraying drones 227
    agriculture spraying flight planning 229
    navigation and precision 228
    payload capacity 228
    precautions and safety 229, 230
    spraying mechanisms 228
    spray tank integration with drone 228, 229
    Variable Rate Technology (VRT) 228
ailerons 32
airframe 14, 238
    designing and developing 67, 68
    selection considerations 238
    subcomponents 15-17
all-up weight/gross weight 30
altitude estimate graph 204
Analog-to-Digital Converter
    (ADC) drivers 111

Index

angular acceleration FFT graph 207
anti-clockwise yaw movement 40
ArduPilot
  assembling 251, 252
  wiring 251, 252
ArduPilot firmware 71
  URL 121
ArduPilot flight stack 112
ArduPilot flight stack, architecture 113, 114
  communications layer 114
  external sensors 117
  flight code 115, 116
  operating system 116
  UI/API 114
arms 16
assisted modes 149
attitude estimate graph 204
attitude hold mode 150
autonomous mission planning 142
autonomous modes 151
  hold mode 151
  land mode 152
  mission mode 151
  return mode 151
  take-off mode 152
autonomous waypoint mission
  planning and executing 157-165
avionics
  configuring 185
  firmware, flashing 185
  flight controller, configuring 186-188
  setting up 185

B

barometer 86
  working 86

Basic Input/Output System (BIOS) 92
Battery Eliminator Circuits (BECs) 233
beyond VLOS (BVLOS) 22
bill of materials (BOM) 171
brain of drone 26
brushless DC (BLDC) motors 19, 61
  KV ratings 61, 62
  operating voltage 62
  torque 62

C

camera/payload
  selection considerations 249
CAN (controlled area network) 78
change detection analysis 227
Chibi Operating System (ChibiOS) 116
ChibiOS, in ArduPilot
  role 116
clockwise yaw movement 39
cloud-based GCS 120
  examples 120
command and control system 22
  components 22-25
comma-separated values (CSV) 199
communicating devices
  used, for integrated GCS system 75, 76
communication protocol 72
communication system 73
  parameters 73, 74
control logs 200
control system development
  electronics speed controller (ESC) 68, 69
  flight controller (FC) 70
CPPM 53
cruise 31

cruise speed 31
cruising in three-dimensional (3D) space 31
current measurement 88
   uses 89
current ratings, for ESC
   burst current 69
   continuous current 69

D

desktop-based GCS software 47
Digital Elevation Models (DEMs) 226
Digital Surface Models (DSMs) 225, 226
Directorate General of Civil
   Aviation (DGCA) 13
DJI Naza 73
Doodle Labs radio 75
   applications 76
drag 31
drone
   airframe 14-17
   airframe, assembling 173, 174
   avionics system 17
   battery, installing 181, 182
   command and control system 22-27
   components, assembling 173
   connecting, methods 134-139
   defense use cases 6
   drive train 18
   ESCs, assembling 177, 178
   ESCs, configuring 177, 178
   evolution 4
   flight controller, installing 178-180
   history 4
   mechanical and structural components 14
   motor, assembling 175-177
   payloads, integrating with 220, 221
   PDB, mounting 180, 181

   power system 20, 21
   propulsion system 18
   RC transmitter and receiver, configuring 183
   subsystems 17
   telemetry connection 183
   tools and components 171
drone configurations 253-260
   ESC calibration 261, 262
   MOTOR TEST window 260
DroneDeploy 225
   URL 225
drone ecosystem
   safety regulations 232
drone firmware 98
   structure 99-101
   versus general firmware 98
drone flight logs 198, 199
   application 198
   issues, analyzing 198, 199
   types 200
drone flight stacks 101
   ArduPilot flight stack 112
   basic PX4 controller loop diagram 105-107
   closed source 102-105
   closed source, examples 103
   open source 101
   open source, examples 102
   PX4 flight stack 107-112
drone's components
   airframe 238
   camera/payload selection considerations 249
   flight stack selection 247
   GPS module selection 248
   propulsion system/power trail 239
   selecting 237
   telemetry system selection criteria 250, 251
drones, types and specifications
   fixed-wing drone 11

fixed-wing VTOL/hybrid drone 12
multirotor 7-11
tilt-rotor drone 12, 13
drone system, conceptualizing 58
application 58
endurance 58
features and functionalities 58
max payload weight 58
range 58
Drotek DP0601 78

E

eight motors (in Quad)/octa-quad 10
eight motors/octocopter 9
Electronic Speed Controllers (ESCs) 42, 61, 68, 100, 170
selecting, consideration 69, 70
working 68
Engine Control Units (ECUs) 93
enterprise GCS 121
examples 121
environmental assessment data 227
error logs 200
ESC calibration window 261, 262
ESC selection 244, 245
Extended Kalman Filter 2 (EKF2) 105

F

failsafe
configuring 190-192
setting up 190-192
Fast Fourier Transform (FFT) graph 207
Fast Real-Time Publish-Subscribe (FastRTPS) 109
firmware 71, 92
components 93, 94

example 95
roles 93, 94
tools, used for developing 95-97
types 92, 93
versus software 97, 98
Firmware-over-the-Air (FOTA) 94
first-person view (FPV) drones 34
fixed-wing drone 11
fixed-wing VTOL/hybrid drone 12
flight axis 31
flight controller (FC) 51, 68-70
selecting 70-72
flight control systems (FCSs) 197
flight mode 148
assigning, in GCS 153
assigning, via remote control 152, 153
modifying, in GCS 153
flight modes, of remote controller switch
configuring 154-157
flight modes, types 148, 149
autonomous modes 151
manual mode/stabilize mode 149
semi-autonomous modes 149
flight physics
quadcopter 34-37
flight stack selection 247
flight time
calculating 67
calculating, formula 67
FlySky RC 49
four motors/quadcopter 8

G

GCS hardware 42
components 42
GCS software 42, 46
components/windows 43

download logs, analyzing 202, 203
graphs, interpreting 203
logs, downloading from flight
  controller with QGC 201, 202
working, on logs and graphs 201
GCS software, types 43
  handheld GCS 45
  large vehicle-mounted GCS 43
  non-vehicle-mounted portable GCS 44
geographic information systems (GISs) 236
geospatial survey 236
  usage, in industry 236
**global navigation satellite
    systems (GNSS) 76**
Global Positioning System (GPS) 16, 25, 87
  benefits 88
  communication protocol 77
  selection 76
  using, in drones 88
  working 87, 88
GNSS navigation device
  power consumption 77
  size 77
  weight 77
GNSS satellite constellations 76
  comparing 77
GPS module
  selection considerations 248
Grand Unified Bootloader (GRUB) 92
Graphics Processing Units (GPUs) 93
graphs
  actuator controls FFT graph 208, 209
  actuator estimate graph 207
  altitude estimate graph 204
  angular acceleration FFT graph 207
  attitude estimate graph 204
  local position estimate graph 206

power graph 211
power spectral density (PSD) 210
ground control software 119
**ground control station (GCS) 22, 41,
    42, 91, 119, 120, 147, 197**
  components 42
  flight mode, assigning 153
  flight mode, modifying 153
ground data terminal (GDT) 42
Ground Sampling Distance (GSD) 223
Gyroscope 82-84

H

handheld GCS 45
Hardware Abstraction Layer (HAL) 114
Here3 78
Herelink 75
high altitude long endurance (HALE) 58
high TWR 61
hold mode 151
horizontal movements 38
hover motion 37, 38
hybrid aircraft 12

I

image stitching 224
  Agisoft Metashape 225
  DroneDeploy 225
  Pix4D 225
IMU components
  accelerometers 82
  gyroscopes 82, 83
  magnetometers 82, 84

IMUs, usage in unmanned systems
   in advanced flight controller 86
   redundant IMU setup 85
   redundant IMU setup, features 86
industry
   geospatial survey, usage 236
industry-famous flight controller boards
   DJI Naza 73
   KK flight controller 72
   Pixhawk 2.8 flight controller 72, 73
   Pixhawk Cube Orange 73
Inertial Measurement Unit (IMU) 78, 81, 99
   composition 82
Integrated Development Environment (IDE) 95
integrated GCS system
   with communicating devices 75, 76
Interrupt Service Routines (ISRs) 94

K

KK flight controller 72

L

land mode 152
large drones 13
large vehicle-mounted GCS 43
lift 31
Light Detection and Ranging (LiDAR) 4, 219
line of sight (LOS) 7
Lithium-Polymer (LiPo) 170, 243
local position estimate graph 206
Long Range Systems (LRS) 184
long-range telemetry components
   mounting 184
Long-Term Evolution (LTE) 48
LoRaWan 48
low TWR 61

M

magnetometer 82, 84, 87
maiden flight
   performing 194, 195
   setting up 192, 193
manned systems 4
manual mode/stabilize mode 149
maximum take-off weight (MTOW) 30, 58
mechanical airframe
   features 17
medium altitude and long endurance (MALE) 58
medium drones 13
MEMS Accelerometer devices
   examples 83
Message Queue Telemetry Transport (MQTT) 48
Micro Aerial Vehicle 109
Micro Air Vehicle Link (MAVLink) 47
microcontroller 70
micro drones 13
mission logs 200
mission mode 151
Mission Planner/APM planner 121
   download link 121
   features 122, 123
mobile GCS apps 121
   examples 121
mobile/tablet-based GCS software 47
moment 35
mono 121

motor datasheet
    technical qualifications 240
motors
    characteristics 61, 62
    selection 65
    selection considerations 240
    specifications 62-64
    thrust charts 62
MOTOR TEST tab 260
multirotor 7
    12 motors (in Hexa)/deca-hexa copter 11
    eight motors (in Quad)/octa-quad 10
    eight motors/octocopter 9
    four motors/quadcopter 8
    six motors/hexacopter 9
    three motors/tri-copter 8
    two motors/bi-copter 7

N

nano drones 13
navigation system
    enabling, components 25, 26
    selection 76
non-vehicle-mounted portable GCS 44
Non-Volatile Memory (NVM) 92

O

open-LTE 48
open source GCS 120
    examples 120
operating input voltage 70
Operating System (OS) 92
orthomosaic maps 225

P

payload release mechanism 230
    consideration, parameters 231, 232
    types 230, 231
    working 231
payloads 26
    integrating, with drone 220, 221
    used, in aerial survey missions 219, 220
PDB selection 245, 246
Personal Protective Equipment (PPE) 230
PID controllers 211
    methods 214
    role 213
    tuning 214-216
    uses 211, 212
pitch angle 35
pitch movement/forward-backward
        movement 32, 39, 40, 51
Pix4D
    URL 225
Pixhawk
    implementation within 233
Pixhawk 2.8 flight controller 72
Pixhawk Cube Orange 73
point clouds 227
position mode 150
Post-Processed Kinematics (PPK) 222
post-processing 224
power distribution board (PDB) 174
power graph 211
power spectral density (PSD) 210
power system 20
    components 20, 21
PPMSUM 53

propeller
  material 242
  selecting 65, 241
  selecting, considerations 241
Proportional-Integral-Derivative (PID) 99, 170, 197
proprietary GCS 120
  examples 120
propulsion system 18, 239
  airframe dimensions 243
  autopilot selection 246, 247
  battery selection 66, 243, 244
  components 19
  designing 59
  ESC selection 244, 245
  flight time, calculating 67
  motors, characteristics 61, 62
  motor selection 240
  PDB selection 245, 246
  power and efficiency 243
  selection criteria, for motor and propeller combination 60
  thrust-to-weight ratio (TWR) 60
pulse position module (PPM) 53
Pulse Width Modulation (PWM) 54, 68, 70, 101, 153, 229
PX4 controller loop diagram 105
  stages 105-107
PX4 firmware 71
PX4 flight stack 107-112
  examples 111

Q

QGroundControl (QGC) 123, 198
  download link 123
  features 123

quadcopter
  flight physics 34-37

R

racing drones 34
RC controller
  working, modes 50, 51
RC protocols 51, 52
RC receiver 49
  mounting 183
RC transmitter 49
  mounting 183
RC transmitter system
  components 49
Real-Time Kinematics (RTK) 78, 222
real-time operating systems (RTOSs) 116
receiver (RX) protocols 51-53
  pulse position module (PPM) 53
  pulse width modulation (PWM) 54
  serial protocol 55
redundant power systems 233
redundant sensors 232
remote control (RC) 22, 41, 48
  flight mode, assigning via 152, 153
remotely piloted aircraft systems (RPAS) 5, 129, 133
return mode 151
return-to-home (RTH) 186
Return to Launch (RTL) 25
RFD 900 radio modem 75
roll angle 35
roll movement 32, 51
RPA/drone configuration 139
  failsafes 140-142
  parameters 140

S

SBUS protocol 55
semi-autonomous modes 149
  attitude hold mode 150
  position mode 150
sensor calibration 188-190
sensor fusion 89
sensors 71
serial bus (SBUS) 48
serial protocol 55
simulation GCS 121
  examples 121
six motors/hexacopter 9
SIYI AK28 75
skeleton 14
small drones 13
software
  versus firmware 97, 98
Software Development Kit (SDK) 103
Solid-State Drives (SSDs) 93
source code availability 102
state estimation 89
Structure-from-Motion (SfM) 225
survey-based drones 218
  aerial surveys 218
survey drone requirements
  all-up weight/gross weight 237
  autonomous drone flight planning 237
  durability and weather resistance 237
  GPS 236
  high-quality imaging sensors 236
  long flight time 237
  payload capacity 237
  real-time monitoring and telemetry 237
  setting up 236
survey planning 144
SWaP (size, weight, and power) 75

Synerex MDU-2000 78
system redundancy 232
  implementation within 233

T

take-off mode 152
Team Access Management 128
telemetry connection 183
  long-range telemetry components, mounting 184
  long-range telemetry system, configuring 184, 185
  long-range telemetry system, connecting 184, 185
  protocol, selecting 184
telemetry logs 200
telemetry protocol 47
  examples 47
telemetry system
  selection considerations 250, 251
three motors/tri-copter 8
throttle 33, 51
thrust 30
thrust charts 62-64
thrust-to-weight ratio (TWR) 34, 60
  calculating 60
  for general purpose 60
  importance 60
  requirements 61
tilt-rotor drone 12
transmitter (TX) protocols 51, 52
triple redundant IMU setup
  algorithm 85
  configuration 85
  redundancy 85

two motors/bi-copter 7
types, drone flight logs
   control logs 200
   error logs 200
   flight logs 200
   mission logs 200
   telemetry logs 200

U

UgCS 124, 127
   download link 124
   features 125
Unified Extensible Firmware Interface (UEFI) 92
unmanned aerial system (UAS) 5
Unmanned Aerial Vehicles (UAVs) 3, 4, 98, 173, 227
   applications 5, 6
   system composition 13, 14
unmanned ground vehicles (UGVs) 4
unmanned system
   need for 5
unmanned vehicles 4
unmanned water vehicles (UWVs) 4
User Datagram Protocol (UDP) 138

V

Variable Rate Technology (VRT) 228
vegetation indices 227
visual LOS (VLOS) 22
voltage measurement 89
   uses 89

W

waypoint planning 143
   mission 143
weight 30
   all-up weight or gross weight 30
   maximum gross take-off weight 30
wheelbase 68

Y

yaw angle 35
yaw movement 33, 51

Z

Zigbee 48

www.packtpub.com

Subscribe to our online digital library for full access to over 7,000 books and videos, as well as industry leading tools to help you plan your personal development and advance your career. For more information, please visit our website.

Why subscribe?

- Spend less time learning and more time coding with practical eBooks and Videos from over 4,000 industry professionals
- Improve your learning with Skill Plans built especially for you
- Get a free eBook or video every month
- Fully searchable for easy access to vital information
- Copy and paste, print, and bookmark content

Did you know that Packt offers eBook versions of every book published, with PDF and ePub files available? You can upgrade to the eBook version at packtpub.com and as a print book customer, you are entitled to a discount on the eBook copy. Get in touch with us at customercare@packtpub.com for more details.

At www.packtpub.com, you can also read a collection of free technical articles, sign up for a range of free newsletters, and receive exclusive discounts and offers on Packt books and eBooks.

Other Books You May Enjoy

If you enjoyed this book, you may be interested in these other books by Packt:

Practical Arduino Robotics

Lukas Kaul

ISBN: 978-1-80461-317-7

- Understand and use the various interfaces of an Arduino board
- Write the code to communicate with your sensors and motors
- Implement and tune methods for sensor signal processing
- Understand and implement state machines that control your robot
- Implement feedback control to create impressive robot capabilities
- Integrate hardware and software components into a reliable robotic system
- Tune, debug, and improve Arduino-based robots systematically

Robotics at Home with Raspberry Pi Pico

Danny Staple

ISBN: 978-1-80324-607-9

- Interface Raspberry Pi Pico with motors to move parts
- Design in 3D CAD with Free CAD
- Build a simple robot and extend it for more complex projects
- Interface Raspberry Pi Pico with sensors and Bluetooth BLE
- Visualize robot data with Matplotlib
- Gain an understanding of robotics algorithms on Pico for smart behavior

Packt is searching for authors like you

If you're interested in becoming an author for Packt, please visit `authors.packtpub.com` and apply today. We have worked with thousands of developers and tech professionals, just like you, to help them share their insight with the global tech community. You can make a general application, apply for a specific hot topic that we are recruiting an author for, or submit your own idea.

Share your thoughts

Now you've finished *Drone Development From Concept to Flight*, we'd love to hear your thoughts! Scan the QR code below to go straight to the Amazon review page for this book and share your feedback or leave a review on the site that you purchased it from.

`https://packt.link/r/1837633002`

Your review is important to us and the tech community and will help us make sure we're delivering excellent quality content.

Download a free PDF copy of this book

Thanks for purchasing this book!

Do you like to read on the go but are unable to carry your print books everywhere?

Is your eBook purchase not compatible with the device of your choice?

Don't worry, now with every Packt book you get a DRM-free PDF version of that book at no cost.

Read anywhere, any place, on any device. Search, copy, and paste code from your favorite technical books directly into your application.

The perks don't stop there, you can get exclusive access to discounts, newsletters, and great free content in your inbox daily

Follow these simple steps to get the benefits:

1. Scan the QR code or visit the link below

https://packt.link/free-ebook/9781837633005

2. Submit your proof of purchase
3. That's it! We'll send your free PDF and other benefits to your email directly

www.ingramcontent.com/pod-product-compliance
Lightning Source LLC
Chambersburg PA
CBHW080803020526
44114CB00046B/2764